Anonymous

The Practical Use of Meteorological Reports and Weather-Maps

Anonymous

The Practical Use of Meteorological Reports and Weather-Maps

ISBN/EAN: 9783337337087

Printed in Europe, USA, Canada, Australia, Japan

Cover: Foto ©berggeist007 / pixelio.de

More available books at **www.hansebooks.com**

WAR DEPARTMENT.

CIRCULAR.

THE PRACTICAL USE OF METEOROLOGICAL REPORTS

AND

WEATHER-MAPS.

OFFICE OF THE CHIEF SIGNAL OFFICER,

DIVISION OF TELEGRAMS AND REPORTS FOR THE BENEFIT OF COMMERCE.

WASHINGTON:
GOVERNMENT PRINTING OFFICE.
1871.

WAR DEPARTMENT,

OFFICE OF CHIEF SIGNAL OFFICER,

DIVISION OF TELEGRAMS AND REPORTS FOR THE BENEFIT OF COMMERCE.

Washington, D. C., Nov. 1, 1871.

CIRCULAR.

The following paper is published by direction of the Secretary of War.

It is the object of the publication to put it in the power of the largest number to make use of, and to profit by, the labors of this Office; to enable them to test, and to avail themselves of, some of the laws and generalizations by which meteorologists are guided; and to afford the means by which at once to supplement, judge of, and aid the work of the Department.

Albert J. Myer

Brig. Gen'l and Chief Signal Officer, U. S. A.

THE PRACTICAL USE OF METEOROLOGICAL REPORTS AND WEATHER-MAPS.

In pursuance of the duty imposed upon the Secretary of War by the law providing for the announcement by telegraph and signal of the approach and force of storms, and under his direction, the office of the Chief Signal Officer of the Army, at the War Department, causes meteorological observations and reports to be made, daily and nightly, at 55 stations.

The office Division of Telegrams and Reports for the Benefit of Commerce is organized for the preparation, receipt, and use of these reports.

At every station three observations are taken daily, at the same moment of actual (not local) time for all stations, by the Observer Sergeants of the Signal Service. The reports are immediately telegraphed to the office of the Chief Signal Officer at Washington.

By a carefully arranged system of telegraphic operation, copies of the full reports of all stations thus transmitted to Washington, or of portions of them, are sent at the same time to many of the Signal Service stations in principal cities and towns.

At each station so receiving a tabular report, one or more Bulletins are published. The observations are made synchronously at the different stations, at the exact hours, 7.35 A. M., 4.35 P. M., and 11.35 P. M., Washington time.

The full reports from all stations are telegraphed to, and received at, Washington, translated from cipher and published in the form of Bulletins of Reports by the hours of 9 A. M., 6 P. M., and 1 A. M., respectively, (Washington time.) The Bulletins of Reports are designated as follows: That published at 9 A. M., the "Morning Report;" that published at 6 P. M., the "Afternoon Report," and that published at 1 A. M., the "Midnight Report." The Bulletins, wherever published, at Washington or elsewhere, exhibit the following particulars, viz: Height of Barometer; Change since last report; Thermometer; Change in last 24 hours; Relative Humidity, in per cent.; Direction of

Wind: Velocity of Wind, in miles per hour ; Pressure of Wind, in pounds per square foot : Force of Wind; Amount of Cloud ; Rainfall since last report, in inches and hundredths, and State of Weather.

The following places, now occupied as Stations of this Division, were selected as of most immediate importance for meteorological purposes, and possessing telegraphic facilities. Those which are gradually to be added during this year are already designated by the Secretary of War, for similar reasons, with a design to perfect the net-work of the system so far as the appropriations allow:

Portland, Me.	* Oswego, N. Y.	Escanaba, Mich.
Boston, Mass.	* Rochester, N. Y.	Marquette, Mich.
* New London, Conn.	* Buffalo. N. Y.	* Davenport, Iowa.
* New York City, N. Y.	* Cleveland, Ohio.	* Leavenworth, Kansas.
Philadelphia, Pa.	* Toledo, Ohio.	* Cairo, Ill.
Baltimore, Md.	* Detroit, Mich.	* Cape May. N. J.
* Washington, D. C.	* Chicago. Ill.	Galveston, Texas.
* Wilmington, N. C.	* Milwaukee, Wis.	Montreal, Canada.
Charleston, S. C.	* St. Paul, Minn.	Punta Rassa, Fla.
Savannah. Ga.	* Du Luth, Minn.	* Memphis, Tenn.
Augusta, Ga.	* Pittsburg, Pa.	* Nashville, Tenn.
Lake City, Fla.	Knoxville, Tenn.	* Cincinnati, Ohio.
Key West, Fla.	Indianapolis, Ind.	* St. Louis, Mo.
Mobile, Ala.	Lynchburg, Va.	Omaha, Neb.
New Orleans, La.	Burlington, Vt.	Cheyenne, Wy. T.
San Francisco, Cal.	* Keokuk, Iowa.	Corinne, Utah.
Norfolk, Va.	* Grand Haven, Mich.	Shreveport, La.
Mt. Washington, N. H.	* Vicksburg, Miss.	Louisville, Ky.
Jacksonville, Fla.		

At those places marked with a star, the number of reports received and published in the Bulletin is sufficiently large to permit them to be used in the manner pointed out in this paper. The remaining stations, either on account of special reasons connected with their locality, or of those relating to the telegraphic system, are now chiefly used as reporting stations.

The morning and afternoon reports (Bulletins) are posted at each of the local Signal Service offices, and at a number of other public places in the cities and towns to which they are transmitted.

They are always open for examination. At the more prominent stations, and those in principal cities, large Weather-Maps are also posted every morning, exhibiting, by means of changeable symbols, the reports of the morning observations at the different stations. The midnight report (Bulletin) is gratuitously furnished to every morning newspaper published in the city at which a station of observation may be, which will insert it in its columns. The morning report is also delivered to afternoon papers in time for publication.

The Observers at each station are instructed to afford every facility to the press and to the public for the earliest receipt and most extended use of the reports and information at their respective offices.

In addition to the Bulletins, a statement of Synopses and Probabilities is prepared at the office of the Chief Signal Officer, and thence issued thrice daily. It is immediately furnished to the Associated Press, by which it is telegraphed to all its agencies throughout the country.

The Synopses and Probabilities, with which the public is familiar through the columns of the different newspapers, are issued from the office of the Chief Signal Officer at 1 A. M., 10 A. M., and 7 P. M., daily.

In the study of local Probabilities, the student should make sure that he has before him (as in the columns of the local newspapers) the latest Synopses and Probabilities issued at Washington. To be sure of such facts, he must notice the hours at which they are dated from the Office in Washington. *The Midnight Reports, dated at 1 A. M. of each day, ought to be found in the morning newspapers of that day. The Morning Report, dated at 10 A. M. of each day, is furnished in time for the afternoon and evening papers.*

At places where stations are established, the use of the bulletined reports, in the mode suggested in this paper, would often, perhaps, enable the student to make forecasts of the weather with greater local particularity than can be expressed in the "Synopses and Probabilities" telegraphed to the press, as the latter must, in a limited number of words, give generalizations embracing the whole country; and it is believed that, in many places distant from any station, but on or near the lines of railways or steamers, or with other modes of rapid communication, the Bulletins can be utilized in a corresponding manner. IF, AT THE STATION NEAREST TO PERSONS INTERESTED, AS, FOR INSTANCE, THE BOARD OF TRADE, OR CHAMBER OF COMMERCE, OR THE METEOROLOGICAL COMMITTEE OF AN AGRICULTURAL SOCIETY, OR INDIVIDUALS INTERESTING THEMSELVES IN THE STUDY OF PRACTICAL METEOROLOGY, NO NEWSPAPER PRINTS THE BULLETINS FURNISHED GRATUITOUSLY BY THIS OFFICE FOR THAT PURPOSE, THERE IS A STRONG

PROBABILITY THAT, UPON PROPER APPLICATION MADE TO THE EDITORS OR PROPRIETORS, THEY WOULD BE PRINTED, AS OF INTEREST TO THE SUBSCRIBERS AND READERS. In cases where delay would thereby be avoided, arrangements can often be made with the publisher to have copies of the newspaper containing bulletins sent in advance of its delivery by mail. By such means, and others which will suggest themselves, a record of meteorological conditions elsewhere can be obtained in many places within so few hours after the observations are taken at the different stations as to enable a student to make for himself many important deductions. The accuracy of these would be greatly assisted by local observations made at the same time as those of the Observers of this Division, with similar instruments, and by frequent local observations made during any time at which there is especial interest or anxiety as to the probable weather.

The navigator, the agriculturist, or the student can supplement in this way, by the readings of his own instruments and his local knowledge, the reports and information furnished by the United States, and is fitted to arrive at intelligent conclusions as to the data before him.

There is as yet no provision for furnishing instruments, on the part of the United States, to any other than the Observer-Sergeants, although it is possible that some such authority may hereafter be given. Full information as to the instruments and their use will, however, be furnished to all persons who may provide them for themselves.

In addition to the Weather Bulletins and the " Synopses and Probabilities," a graphic weather chart or map is issued thrice daily from the office of the Chief Signal Officer of the Army, at the War Department. To those who know how to use them, all of these republications offer valuable help in estimating the probable character of the weather at any station, or over any district, during the following day, and often for a still longer period. The bulletins and graphic charts, properly filled, convey the same information, with this difference : while the former merely tabulates the reports alphabetically, the latter reveals to a single glance of the eye a synoptic view, at once, of the meteoric conditions at the different stations, and of the deductions thence to be made as to the conditions of the atmosphere then extending over the continent.

The graphic charts are of additional value, from the fact that it is often possible to trace upon them, in lines, the progress of storms, or the change of meteoric condition (as the movement of an area of high or low barometer) from report to report, and thus, by considering the past, and by applying laws and generalizations reasonably well established, to estimate more easily the " Probability" of the future.

Those who receive the Bulletin through the newspapers, or otherwise, can, if they desire, so easily transfer its information to the blank charts, that the reader is supposed to be in possession of, or familiar with, a series of such weather maps.

As the anticipation of dangerous winds is at present the most important subject for examination, the following remarks are made with the view of directing the attention to those matters that are considered to have a direct bearing upon the subject of storms and gales in the United States. The study of the science of meteorology and of the theories that have been formed to explain the complicated phenomena that are daily presented in our atmosphere, will be but lightly touched upon, since persons of education are already in possession of sufficient inductive generalizations to give a fair insight into the actual nature of the changes that are taking place at every hour of the day.[*]

ABBREVIATIONS USED IN THE PRESS REPORTS.

It may be well to state here, that in the Weather Synopses and Probabilities, emanating from the Signal Office, different parts of the country are thus designated:

Maine, New Hampshire, Vermont, Massachusetts, Connecticut, and Rhode Island, are alluded to as the *New England States* or the *Northeast*, or simply as the *Eastern States*.

New York, New Jersey, Pennsylvania, Maryland, District of Columbia, and Virginia as the *Middle States*, or sometimes as the *Middle Atlantic States*.

North Carolina, South Carolina, Georgia, and Northern and Eastern Florida, as the *South Atlantic States*.

Western Florida, Alabama, Mississippi, Louisiana, and Texas, as the *Gulf States*.

Sometimes the Gulf States, the South Atlantic, Virginia,

---

[*] Blank charts on which to enter observations can be obtained from the Office of the Chief Signal Officer, at Washington, D. C., at actual cost, ($2 75 per hundred,) and will be sent by mail, free of other expense, to applicants.

Tenuessee, Kentucky, and Arkansas are grouped together as the *Southern States*.

The Lower Lakes, when used, means Lake Erie and Ontario.

The Upper Lakes are Lake Superior, Huron, and Michigan.

The Northwest, popularly, means the country lying between the Mississippi and Missouri Rivers.

The Southwest means Texas, Indian Territory, and New Mexico.

Pacific Coast or *Pacific States* includes California, Oregon and Washington Territory.

The *Ohio Valley* includes the belt of country about two hundred miles broad from Pittsburg to Cairo.

The *Mississippi Valley* includes a belt of about the same width, from Vicksburg to Davenport.

The extensions "from Missouri to Ohio," &c., &c., refer to areas reaching to and including the central portions of the States named. Thus, a report "Westerly winds extending from Iowa to Pennsylvania," would signify that those winds would be felt in the interior of those States as well as over the territory lying between them of the respective States.

In "*the Coasts*, &c.," is included the land between the coasts and the parallel range of coast hills or mountains. In Texas, Louisiana, and Northern Florida, a belt of land extending a hundred miles inward would be included.

Winds are said to blow from N. E. when they are generally included within the quadrant from N. to E., &c., and similarly for other directions.

THE GRAPHIC WEATHER-MAP.

A weather-map should be made out as often as a Bulletin is received; and it is a very great advantage to be able to do this thrice daily; one map every day is the least than can be usefully studied. It is found best to preserve the maps in books of a hundred or so, they being very exactly superposed upon each other.

Owing to the translucency of the paper on which they are printed, it is generally possible to see through several of the sheets, and thus more easily compare the successive phases of the weather.

The record for each station should be entered in its appropriate place; the wind and weather may be easily indicated by arrows and black or white circles.

All the entries being made with a black lead pencil, the whole operation will barely consume fifteen minutes.

Auxiliary charts of the same size should be at hand, on which are entered the isobarometric* and isothermal lines for the month, and such other data as need to be continually borne in mind. Especially important is it that the eye should be familiar with the ordinary charts of physical geography, on which are pictorially given the belts of trade winds and the anti-trade winds, the mountainous and alluvial regions, the plateaus, &c., and it should be borne in mind that the climatic belt continually moves up and down on the earth's surface, with the sun's apparent annual motion. Relief-maps, or orographic charts showing, pictorially, the face of the country whose meteorology is studied, will add the liveliest interest and clearness to the investigation.

It is probable that elementary charts showing the principal topographical and meteorological features of North America, and on precisely the same scale as the weather-map, will be eventually published by the office of the Chief Signal Officer; but for the present the desired information may be obtained from such works as Keith Johnson's Physical Atlas; Guyot's, Cornell's, or Maury's Physical Atlas and Geography; Buchan's Meteorology and his essay on the Distribution of Atmospheric Pressure; Blodgett's Climatology of the United States; Coffin's Winds of the Northern Hemisphere.

The manner of tracing Isobarometric lines on the *Weather-Map* is as follows:

1. Record with a lead pencil in small plain figures at each station the height of barometer as given in the Bulletin.

2. Note the location of the regions of lowest and highest barometer. Intermediate between these, select such a reading as may frequently occur, and draw a red line through all stations at which it does occur, or may be presumed to occur, as shown by neighboring reports. Draw a similar line, correcting the readings that are all one-tenth inch lower than the first, and again a third line, correcting readings still a tenth lower, and so on until the area of low pressure is completely inclosed. A similar process is to be gone through with the readings on the other side of the first line until the area of highest pressure is inclosed by its isobars.

---

* Isobarometric (sometimes abbreviated to Isobaric) lines (or Isobars) means lines along which the mercurial column indicates equal pressure.

By copying in blue or dotted red, the lines of the previous map, the change in location of the area of lowest pressure since the last report, as also the direction in which the storm (if there be one) is moving, and the rate at which it is progressing, are' readily shown. The following is an illustration: The dotted line showing the Isobaric line of 29.50, located by the readings supposed to have been found upon the map on either side of that line. The specimen of a War Department Weather-Map accompanying this paper will enable this subject to be fully understood:

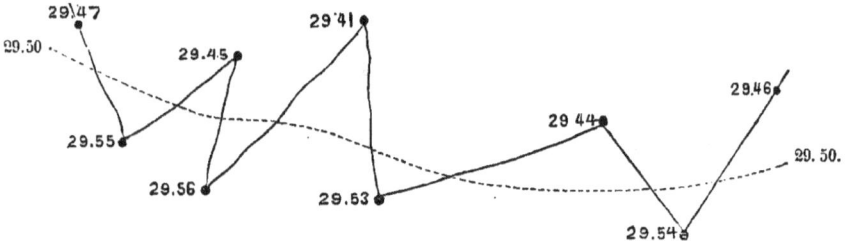

THE ATMOSPHERIC PRESSURE.

The pressure of the atmosphere at a given place is not expressed by the weight of the mercury sustained in the tube of the barometer by that pressure, but by the vertical height of the mercurial column.

In comparing barometric observations at different stations, it is necessary to take into account their respective elevations above the sea-level; for, as there is an ascent above the sea, the heavier strata of the atmosphere are left behind, and the pressure being thereby reduced, the height of the mercury is proportionately reduced.

A slight acquaintance with the Weather-Map will, however, bring to notice the fact that the barometric heights, as reported from Mount Washington, Cheyenne, and other very elevated stations, even after all allowances are made for the altitudes, are very different from those at neighboring stations which are more nearly on the sea-level. Attention is, therefore, for the present confined to the barometric readings at those stations that are less than 2,000 feet above the sea, which includes nearly all those from which reports are at present received.

A few trials will convince one that it is almost always possible to draw a continuous line through those points where the barometer stands at 30.00 inches. If, furthermore, from two neighboring points there are reported, respectively, 29.95 and 30.05, then half way between them, it may, in the majority of cases, be assumed that the barometer stands at 30.00. By con-

WAR DEPARTMENT WEATHER MAP.

SIGNAL SERVICE U.S.A.

DIVISION OF TELEGRAMS AND REPORTS FOR THE BENEFIT OF COMMERCE

Washington, Saturday, August 26, 1871—7.35 A. M.

necting with red lead pencil lines all the points thus deter-mined, that narrow band over which the barometric pressure is uniform is made visible at a glance.

It is generally well understood that the height of the mercury in the barometer tube is a simple and direct measure of the intensity with which the atmosphere is at that moment pressing down upon the basin of the barometer, and upon the neighbor-ing region of the earth; and not only is the pressure downward, but equally so is it exerted upward and horizontally in all directions.

The pressure varies in intensity at any station with the vary-ing temperature, moisture, depth, and motion of the atmos-phere. There is also a slight regular rise culminating between 9 and 10 A. M., and a fall culminating between 3 and 5 P. M., called the diurnal change.

The average height of the barometer at the level of the sea, on the Atlantic coast of the United States, does not vary much from 30.00; on the western plains it rises to 30.2 in the winter. It diminishes as we approach the Arctic regions.

Now draw other red lines showing where the pressure is lower than its average value, (30.00,) or where the barometer stands at 29.90, 29.80, &c., inches. These lines will almost invariably be found to be upon one and the same side of the line of 30.00, and to be approximately parallel to each other. Oftentimes they gradually inclose within their bounds a central area of small extent, and over which the pressure is decidedly lower than anywhere else on the map. If this central area be not too near the limiting line of the Signal Stations, it will be found to be completely inclosed by the encircling lines of equal pressure, or Isobarometric lines, as they are called, words which, for con-venience, are now generally contracted into the simple noun *Isobarics* or *Isobars*.

Similarily, if, on passing to the other side of the line of 30.00 inches of pressure, a system of Isobarics connecting the points where the barometer stands at 30.10, 30.20, &c., be drawn, there will be noticed an area of the highest pressure; though very frequently the stations are not widely enough extended to give the precise limits of this area. In general the areas of average and high pressure are more extensive than those of low pressure.

Could reports be received from the whole of the north tem-perate regions of the earth, from the oceans as well as the con-tinents, it would undoubtedly be found that each area of lowest pressure is completely separated, by areas of average and high pressure, from its neighboring low-pressure areas. For the limited region from which the daily reports are received this is found to be approximately true; so that, for instance, a low barometer over the extreme western plains will be bounded by areas of high barometer on the Pacific Coast and the Mississippi Valley, while another area of low pressure prevails in the New England States. On the borders of the great areas of high barometer, and within the areas of low barometer, occur the smaller areas of low pressure with which cyclones, tornadoes, and thunder-storms are associated. It is the study of these areas, of high and low barometer, that is now of particular interest.

The dimensions of these smaller areas vary from that of a few square miles, as in tornadoes, to that of five hundred miles square, as in extensive cyclones. The largest areas of high barometer of interest to this country, are those of the Tropic of Cancer and the South Atlantic and Pacific Oceans, and (in the winter) that which exists in the interior of the North American continent.

While carefully studying the every-day variations of the barometer, the meteorologist will not forget each day diligently to compare the day's Isobars with the mean monthly Isobars on his isobarometric chart. These differ widely from the Isobars for the year as shown in the accompanying chart.

ISOBAROMETRIC LINES, SHOWING, IN INCHES, MEAN ANNUAL ATMOSPHERIC PRESSURE FOR UNITED STATES.

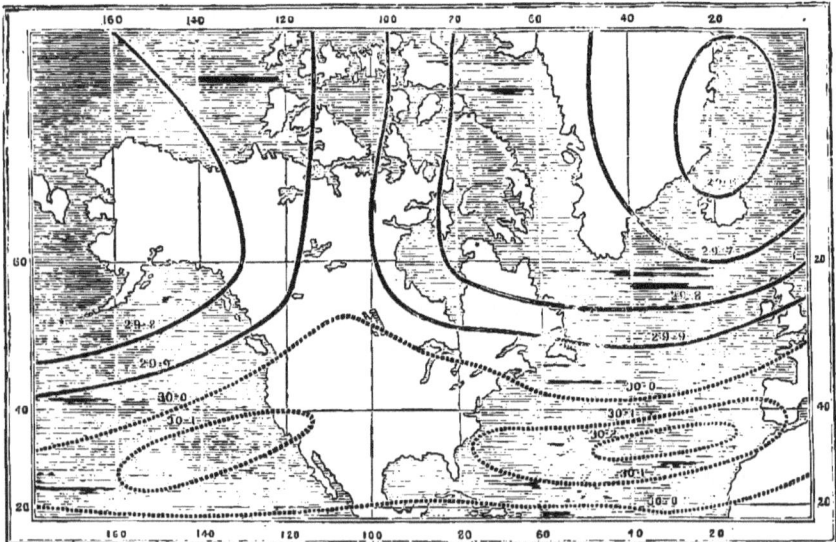

THE WINDS AND THEIR LAW.

Whether considered as the indices, or as the causes of coming changes of weather, no phenomenon is more important than that of the winds. Upon the direction and force of the winds, some meteorologists lay very great stress in every attempt at storm forecasting; and, in order to determine these, it is necessary to draw the Isobars.

Assuming the lines of equal barometric pressure to be drawn on the Weather-Chart, it is at once perceived that, in well-understood accordance with the laws of mechanics, the atmosphere must be pushing from the regions of higher to those of lower pressure. The resulting movement of the air, modified by the forces of inertia and friction, and by the rotation of the earth and local obstructions, is converted into the *local* winds whose directions are indicated by the arrows upon the maps, and whose velocities are given in miles per hour. These winds may be called local winds, as distinguished from the general winds in any section, and from the great currents of air to be hereafter spoken of; the general winds appear to be primarily dependent upon the existence and position of the areas of low and high pressure; the great currents, spreading, as they do, over whole continents and encircling the earth, are largely influenced by, if not entirely dependent upon, the earth's axial rotation.

If the earth were not in rotation on its axis, the winds would uniformly blow in straight lines outward from the center of every area of high barometer toward the surrounding localities of lower barometer. Observation, however, has long since clearly shown, that in this hemisphere, within any area of high pressure, the winds will be found to be not only blowing away from the center, (outwards,) but also to be deflected toward the right hand as they move forward. Observation has also shown, with equal clearness, that in this hemisphere, within any area of low pressure, the winds will blow toward the center, (inwards,) and will also be deflected toward the right hand as they move forward. This deflection to the right has been demonstrated by Mr. Wm. Ferrel, of Cambridge, Mass., to be a mathematical necessity arising from the influence of the earth's diurnal rotation which causes everything moving on its surface to deflect slightly to the right in the northern hemisphere, and to the left in the southern hemisphere. This force, by which, to give a popular illustration, a railroad train is made to bear more heavily on the right-hand rail of the track along which it advances, is the key to the explanation of many phenomena in connection with atmospheric and ocean currents. By considering the influence of this deflection it becomes possible to construct the following table, which shows which winds will generally prevail on each side of areas of high and low pressure:

The observer being—	THE PREVAILING WINDS WILL BE—	
	Low pressure.	High pressure.
On the N. side	N. and E	S. and W.
On the N. W. side	N. W. and N. E	S. E. and S. W.
On the W. side	W. and N	E. and S.
On the S. W. side	S. W. and N. W	N. E. and S. E.
On the S. side	S. and W	N. and E.
On the S. E. side	S. E. and S. W	N. W. and N. E.
On the E. side	E. and S	W. and N.
On the N. E. side	N. E. and S. E	S. W. and N. W.

This same regular distribution of the winds is shown by the arrows in the accompanying diagrams:

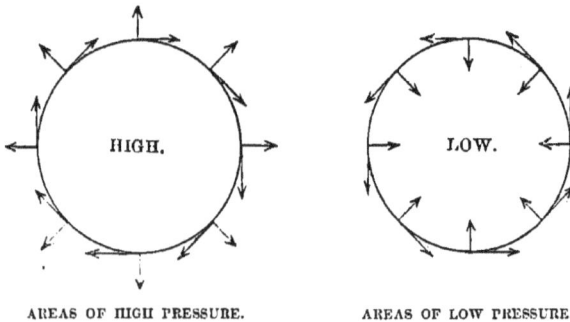

AREAS OF HIGH PRESSURE.　　　AREAS OF LOW PRESSURE.

The deflection from the radial line (*i. e., to the right*) is not always nor usually 90°, as represented in the diagrams, but the angle is generally between 30° and 60°, amounting to 60° or 80° only in case of severe storms, cyclones, &c. In using the above table or diagrams, if, for illustration, it be assumed that the observer is placed within an area of unusually low pressure, then on the northwest side of its center he should find the wind blowing from some point in the horizon between northwest and northeast; if he is within an area of high pressure, and on the northwest side of its center, he should find the wind blowing from some point in the quadrant between southwest and southeast.

When no well-marked central areas are actually within the limits of the map, the winds still unite with the lines of equal pressure in indicating such areas as existing near by. Thus, during the summer, for instance, the southwest and southeast winds of the western plains indicate the low pressure that exists at that season in the Missouri Valley and northward. When the pressure over a considerable area is very uniform, then minor local influences (such as cannot generally be exactly located by the limited number of reports that are at present received) affect the gentle winds that then exist. On such occasions local differences of temperature and moisture, affecting as they do the local pressures, give rise to light winds, extending over only a few square miles, but which still tend to obey the general laws given in the above table.

Vertical as well as horizontal systems of winds, depending upon the disturbances of equilibrium continually taking place in the region of the clouds, always exist in connection with the ordinary horizontal gales; these are, in fact, a most prominent feature of tornadoes and water-spouts.

The system of winds above given is not only that which all observations show to exist, but also that which follows from the mathematical theory of the motions of fluids on the earth's surface, as developed by Ferrel and others, which, as has been already stated, proves that all bodies, whatever be their direction of motion on the earth's surface, in the northern hemisphere are deflected, or trend toward the right hand in their onward progress. This deflection toward the right, small as it is originally, increases with the diameter of the storm, and seems, by its cumulative force, to determine the general direction of the rotation of the West India cyclones, as well as that of the much weaker storms that pass over our continent; in all which it is evident that the air actually has a sinuous spiral motion inward toward the central area of lowest pressure, and at the center upward from the earth's surface; on the outer edge of the tornado and cyclones, a downward current is evident. This same principle (Ferrel's Law) also decides the deflection northward of the afternoon sea-breezes, and southward of the evening land-breezes, as experienced along our

2

Atlantic coast from North Carolina to Florida, which deflection is, however, greatly increased by the general trend of the coast, and the greater friction of air passing over land than over water.

The rule called "Buys Ballot's Law" is not to be confounded with that which, for brevity, is called "Ferrel's Law." The former is expressed as follows:

"If any morning there be a difference between the barometrical readings at any two stations, a wind will blow on that day in the neighborhood of the line joining those stations, which will be inclined to that line at an angle of 90°, or thereabouts, and will have the station where the reading is lowest on its left-hand side."

This rule is a generalization first specially announced for Holland and the neighboring country, and has been found applicable to the weather in Great Britain, in cases where strong winds occur.

The law was first published by Buys Ballot in 1860;* it had also been deduced by Dr. Lloyd in 1854, and had been recognized, under a different form, by the students of tropical cyclones and hurricanes.

This rule will evidently hold good in general, best when the two stations considered are in a line with the center of the neighboring area of high or low pressure. It is, if possible, always best to seek the boundaries of these areas by means of the Isobaric lines, and, having located these, to make use of the laws of mechanics that are applicable to the whole earth's surface. Buys Ballot's rule was originally intended to be applied to regions from which but few isolated reports were received.

Since, however, the centers of disturbance are in continual motion, it is evident that "we must not interpret the rule too strictly." In investigations upon the applicability of this rule to the British weather reports, an allowance of about one hundred miles was made for the movement of the center of disturbance, and thus it was deduced that "94 per cent. of the gradients recorded were succeeded by winds in the direction indicated by the law," and in 62 per cent. both the force and direction were correctly indicated.†

---

* Einige regelen voor aanstaande Weersveranderingen in Nederland voornamelijik id verband met de dagelijksche Telegraphische seinen. Utrecht, 1860.

† See R. H. Scott. An Inqniry into the Connection between Strong Winds and Barometrical Differences. London, 1869.

The force of a local wind at any point, and at any moment, certainly depends primarily upon the relative barometric pressure at points in the vicinity and upon the rapidity with which the pressure has been or at that moment is changing; but the force and direction of the wind at any station are also very materially influenced by the character of the ground in the immediate and distant neighborhood. The wind which on the ocean would blow with a certain velocity, will have but one-half or one-third of that velocity when blowing over hilly country. This is due to the lesser friction on the ocean, and this frictional resistance in two different ways disturbs the direction of wind:

1. If, for example, there is a north wind blowing, very generally over a lake of elliptical shape, such as Lake Michigan, and over the neighboring country, then on the central line of the lake a strong north wind will be experienced, and a feebler one at the points on land far removed from the shore; but at points on the northwest and southeast shores of the lake a north-- west wind will be experienced, while a northeast wind will be observed on the northeast and southwest shores. Similarly, if a south wind blows steadily over the Southern States and coast, it will, to observers on the coast, appear as a southwest wind, and a north wind will be changed into a northeast wind, and this, too, independently of the additional influence exerted by the earth's rotation, which should in this present example increase the extent of those changes, in accordance with the law above given, as first deduced in all its generality by Ferrel.

2. The friction of the earth's surface has a greater influence upon strong than upon feeble winds, and thus does more to retard the tangential than the centripetal motion of the air in the neighborhood of an area of low pressure. Consequently, in severe storms on land the wind is found to be directed more nearly toward the central area of the disturbance than in oceanic storms. Thus in tornadoes the inward and upward motions predominate over the tangential.

Precisely as the velocity over water is greater than over land, so is the velocity far above the earth's surface greater than lower down. Balloon voyages show occasional velocities of one hundred miles per hour. The severest gales on the earth's surface rarely exceed eighty-five miles, though doubtless this has been exceeded in certain tornadoes and momentary gusts, &c. The currents, only a few hundred feet above the earth, have frequently twice the velocity of those observed on the surface, as shown by observations of the velocity of passing cloud shadows.

The friction of air moving over land or water gives rise also to ascending currents, extending over large areas of land, and of which the westerly winds of the Pacific coast and the easterly winds of the Atlantic coast are important examples.

Heavy gales (i. e., those having velocities of forty miles and upward) immediately attend the areas within which the Isobarics are very near to each other, and die away so soon as these lines are seen to separate. In the case of violent but very local storms, the stations will generally fail to give more than a general indication of the disturbance.

The existence of several local winds a short distance above the earth's surface, and on the tops of mountains and in the regions traversed by balloons, and especially over the edges of lakes, the ocean, and arid, dry plains, is very frequently observed, when at the same time neither the barometer nor the wind is affected at the lower stations near by.

This phenomenon is probably in most cases due to the fact that very rapid barometric changes may exist above us, while below there is perfect equilibrium.

Thus suppose that at two neighboring stations the barometers read the same, and the winds are light or calm, but the temperatures are respectively ninety and seventy-five degrees, and these differences occur in summer time at points less than a hundred miles apart. It follows that at an elevation of a mile above these points the barometric pressure may be at least a tenth of an inch different, and this would suffice to set a strong wind in motion.

By considering the influence of moisture in diminishing the density of the air in which it is present, it is shown that the decided differences of pressure that exist at neighboring stations may easily be completely reversed in the strata from one to two miles above the earth.

The destructive power of a wind, or its power to overthrow or move any body, is the difference in the pressure on opposite sides of the body. In steady winds this difference depends not only upon the velocity of the wind, but equally on the shape of the resisting body. Those bodies offer least resistance in which (as in fishes, the hulls of ships, bridge-piers, &c.) the hinder portion receives the backward pressure of the fluid that presses up against it, thus permitting as little approach to a vacuum as possible. In the case of sudden gusts, the resisting body receives the whole force of the impulse precisely as a blow. The atmosphere, though so light, is not devoid of mass and inertia. Air in motion at the rate of one hundred miles per hour strikes obstacles with a force equal to that which the same volume of water would exert if moving at the rate of three and one-half miles hourly.

In descending from higher to lower land, the wind's velocity is affected by the force of gravitation in a manner quite independent of the differences of pressure shown by the horizontal Isobaric lines. Although the primary effect of gravitation is in the vertical direction, yet a portion of this is transferred to the horizontal, and the descending currents acquire increased velocity. Such descending winds are found not only in mountainous countries, but also on the western plains.

Winds of this class are sometimes to be considered as the pushing effect of a heavy descending gas; while ordinary winds are considered as the effects of the drawing action of areas of low pressure. Both cases, however, more properly should be regarded as simple cases of the transfer of air, in the continual struggle going on to maintain that equilibrium of pressure for which the atmosphere is ceaselessly struggling, but which it is not possible for it ever to realize except for brief intervals of time.

THE TEMPERATURE.

The thermometric changes over all parts of the earth's surface are mainly dependent upon the apparent annual and daily motions of the sun, and the grand atmospheric currents.

As fluids and gases are both bad conductors of heat, the distribution of heat in the atmosphere is effected most largely by the winds or by convection, just as in the ocean it is effected by means of the grand aqueous currents.

Although the average temperature is higher at the southern stations than at the northern ones, and higher in the day than at night-time, yet the Weather-Map will disclose innumerable departures from this law, and especially so if any great differences in the pressure or any extended cloudiness exist.

Aqueous vapor visibly suspended in the air, as haze or cloud, serves as an effectual and double shield against the radiation of heat from the earth, and also against the sun's rays themselves. Even the invisible particles of vapor floating in the atmosphere, however rare, present an obstruction to the free passage of heat of low intensity or obscure heat much in the same way as haze and smoke obstruct the light, or as stones in the bed of a water-course retard the flow of that fluid. On the most Alpine situations, where, on account of their loftiness, much less aqueous vapor is interposed between them and the cold stellar regions, radiation is least disputed, and, consequently, when exposed to the direct rays of a serene midday sun the heat is intolerable, while at night the unimpeded radiation produces a corresponding extreme of cold. The temperature observed is the difference between the heat given out and that received in a definite interval of time.

The temperature of the lower air depends primarily, indeed, upon the amount of heat poured down upon the earth by the sun, and the amount absorbed by the air, as the earth radiates its heat back into space; but, in addition to this, the heat held latent in the vapor diffused through the air is at times liberated by the condensation of the vapor into fog, rain, and snow, and then it becomes sensible to the thermometer. During the day a moist atmosphere will become warmer than one that is dry, and during the night the radiation of heat through a moist atmosphere will be less than that through a dry one. During cloudy or hazy weather the radiation is almost wholly cut off, so that a very uniform temperature prevails between the earth and the bottom of the lowest layer of clouds. On the other hand, sufficient heat is absorbed (*i. e.*, becomes latent) in the process of evaporation to materially reduce the temperature of the air; thus it is that " drying winds" are also " cooling." An increase of barometric pressure, by increasing the capacity of the air for moisture, serves to stimulate evaporation, and temporarily reduce the temperature. A diminution of pressure and consequent expansion of confined air produces a lower temperature and diminished capacity for moisture, until the condensing vapor gives out its latent heat.

Again, the lower strata of air receive heat from the upper strata, and radiate back to them, so that the temperature on the earth's surface is in part the result of this interchange. In the normal condition of a clear sky the temperature above should be less than that prevailing below. The abnormal condition is generally the consequence of the elevation of moist air in the regions high above the earth, and the condensation of its moisture consequent upon the expansion of the air. The undue heat thus generated in the upper strata is radiated down to the earth as well as out into space.

Examination of the weather-charts will show that the temperature varies much less over cloudy than over clear districts; that it varies less in low than in elevated regions; that it is warmer on one side of an area of low or high pressure than the other, and generally *warmer in advance of any storm center and colder in the rear.*

The meteorologist, in search of the confines of the storm-area and the path of its advance, will carefully compare the reported temperatures of contiguous stations (lying in this path) both with each other and with the isothermal lines for the season.

By careful attention to the position of the areas of rising and falling temperature, he receives an early intimation of approaching storms, as will be mentioned in a subsequent section.

The relation of the temperature (even for vast regions of country) to the barometric pressure at distant points is full of importance and instruction. For instance, severe frosts and cold have frequently been experienced in Great Britain and Western Europe, traceable directly to an abnormally high pressure of the atmosphere over Iceland, precipitating a powerful polar current of air toward the southeast continuously for periods of two or three weeks.

It is probable that at some future time, weather-telegrams from the West Indian and Sandwich Islands, the North Pacific Ocean, and Alaska, by furnishing barometric readings, may give indications of the weather in the United States.

THE MOISTURE, (RELATIVE HUMIDITY.)

In all localities on the globe, and at all times, moisture, in greater or smaller quantities, exists in the atmosphere, which is, consequently, never absolutely dry. Intervals or interstices occur between the particles of the dry air, which are partially filled with this ever-present aqueous vapor. The more numerous such intervals are, the greater is the *capacity* of the air for moisture; and when these intervals are so full of vapor that the air is incapable of containing or holding any more, it is said to be *saturated.*

An increase of heat increases the capacity of the air for moisture; while, on the contrary, a fall of temperature is the occasion of a corresponding diminution of the capacity for vaporous matter.

The important element of moisture is given in the Signal

Service Bulletins, not in the absolute quantity in which it is found at any given place, but as a percentage of full saturation, or what, in the language of meteorologists, is expressed by the term *Relative Humidity.* This must not be confounded with absolute humidity, which is a very different thing. For, supposing the temperature of the air at a given place to be 40° and fully saturated with aqueous vapor, and then to be suddenly raised to 50° without any addition being made to its store of vapor, its absolute humidity would in each case be exactly the same, but in the former case the weather would, in popular language, be very damp, and in the latter case very dry. In the former case the relative humidity (or *humidity,* as it is often simply called) would be very high, *i. e.*, 100 per cent.; in the latter very low, *i. e.*, 50 per cent.

Watery vapor dissolves in air very much as salt dissolves in water, and as the salt is deposited in crystals whenever the water becomes fully saturated, so, whenever the air becomes fully saturated with vapor, the latter is deposited on the earth in the form of mist, dew, and rain, if the temperature be high, or as frost, hail, or snow crystals if the temperature be low.

One cubic foot of air, having a temperature of 50°, and under a uniform barometric pressure of 30.00 inches, and *fully saturated,* will hold 4.28 grains of water according to Glashier's tables. If, under these conditions, the temperature or the pressure of the air is lowered, there will result a deposition of a portion of the water, and that either in the form of fog, dew, rain, frost, or snow and hail. On the other hand, if there is an increase in the temperature or the pressure, the air becomes capable of holding a larger quantity of vapor, and ceases to be fully saturated. Relative humidity expresses the proportion of vapor actually contained in the air compared with what the air could contain.

By denoting full saturation by 100 per cent., and absolute dryness by zero, the relative moisture of the air at the different stations can be indicated on the map by the proper percentage. [This relative humidity is obtained from the Tables of Relative Humidity, (pp. 59, 60,) where the practical process is fully explained.] This table is directly applicable to such stations as are less than 1,000 feet above the sea. A correction of considerable amount is needed for mountain stations.

The absolute quantity of moisture in the normal condition of the atmosphere decreases with ascent above the earth's surface, but the law of decrease in cloudy and falling weather is, of course, different from that in clear weather. The degree of saturation of the atmosphere increases up to the lower cloud stratum, and rapidly decreases above the top of the highest clouds. Indeed, there is reason to believe that very little moisture enters the higher strata of air, except as it is carried up by the ascending currents; and that rain-falls are mostly derived from low clouds that have derived their moisture from the earth near by.

Fog and dew attend the supersaturation of the lower strata of air. Rain and snow are merely the moisture deposited from supersaturated strata above. The Weather Chart will show an increase of moisture near bodies of warm water, fields of snow, extensive forests and meadows, &c., as compared with dry plains and rocky mountains. The humidity will be found large in advance of storm centers, and small in their rear. It will be greater over warm, cloudy districts, than where cold and clear weather prevails.

Certain winds will be found to be moister than others. The west and northwest are generally the driest in the Mississippi Valley. Dry air almost always predominates on the leeward side of mountain chains, and is the characteristic of the plains and plateaus west of the Mississippi Valley. Dryness will be found attending clearing-up weather. Dampness or a large increase of relative humidity accompanies threatening weather as an almost invariable premonition. Ascending currents of air also increase in dampness; descending currents grow drier.

The smoky haze which spreads to a great distance when extensive forest fires prevail, is composed of minute atoms of charcoal which possess the singular property of attracting moisture to themselves and thus perpetuating dry weather.

THE CLOUDS AND THEIR INDICATIONS.

By entering graphically on the map the general features of the weather and sky, we complete the detailed representation of the atmospheric condition. The clouds by their kinds and changes are indices to the relative temperature, moisture, and pressure existing at high altitudes; by their motions they indicate the nature of the prevailing current of air, showing whether it is from the tropics and hence likely to be warm, or from the polar regions, and cool.

The ascent of expanding warm air gives rise to the *cumulus* clouds, whose flat bases are all on a pretty uniform level. These subside and dissolve when they cease to be fed by rising currents of moist air: the thickness of the cumuli from base to peak is less in cold dry weather than on warm moist days. The *cirrus* clouds are probably formed independently by the radiation of heat outward into the highest regions of the atmosphere, in which case they are composed of snow-flakes, or of spiculæ of ice; and they are also formed of the remnants of the storm-clouds, in which case they are generally composed of warmer vapor. The strong winds that attend areas of low barometer give rise, through the influence of friction, &c., as before stated, to ascending strata of moist air, in which, by expansion or cooling, as the case may be, are produced the *scud* and rain cloud of which there is a fine example in the easterly rains of the Atlantic coast. This scud-cloud, which is at first like a cumulus of irregular shape, subsequently spreads into broad sheets of *stratus* and *nimbus*.

Two or more layers of clouds almost invariably coexist wherever extended rain-storms prevail, the upper layer stretching far in advance of the lower, but descending and merging into the lower over the area on which rain is falling most abundantly. In the rear of this area cumulus clouds are abundant. A general survey of the map will show that cumuli or the cirri first mentioned in the preceding sentence are not inconsistent with fair and clear weather, as these terms are popularly used. An increased accumulation of large cumulus clouds may become cloudy weather, but does not generally presage the extended storms of winter. The cirrus of the second class, sometimes called cirro-stratus, almost always precedes at some distance any extensive rain-storm, whether of winter or summer. The stratus will generally be found to be reported in connection with threatening weather at the different stations.

The classification of clouds into cumulus, cirrus, &c., as originally given by Howard, is indicated on the accompanying plate.

GENERAL ATMOSPHERIC CURRENTS.

Without undertaking in these suggestions to discuss the general system of circulation of the atmosphere, it is necessary to call to mind a few general facts. The trade-winds on the surface of the earth, as is well known, steadily blow from the northeast and southeast in the respective hemispheres toward the meteorological torrid zone, which is a narrow belt where calms and rains prevail at all seasons, and the uniformity of these winds is only disturbed, as in the Indian Ocean, by the

1. Cirrus. 3. Cirro-stratus. 5. Cumulus. 7. Stratus.
2. Cirro-cumulus. 4. Cumulo-stratus. 6. Nimbus.

unequal distribution of land and water in the two hemispheres. After reaching the equatorial regions as surface winds, the air must ascend to a great elevation, and thence move toward the polar regions in high upper currents, which descend to the earth in the region of the polar calms. It is probable that in this way are produced the northerly polar currents on the surface of the earth in high latitudes, but in the middle latitudes, in obedience to mechanical law, westerly winds prevail, which are known as the anti-trades.

In the Temperate Zone of the Northern Hemisphere the most frequent winds for eight or nine months are from the west or southwest, and allowing for the strength of the respective winds, the atmosphere is during the whole year carried to the eastward. The great currents that circulate around continental areas of high or low barometer interfere with and even reverse this eastward motion in the temperate and westward motion in the Tropical Zone, but in the United States, north of the Gulf, any westward motion of the lower winds (and especially in the autumn, winter, and spring months) is to be considered as the result of a local disturbance, which may originate on the immediate surface of the ground, but frequently originates in the lower stratum of clouds. It is this disturbance that induces the surface winds from the northeast and southeast which blow nearly toward the storm center, while west winds prevail far above, and also on the western side. On the east side of a winter storm, and not far from its center, these easterly surface winds may extend for two miles upward above the earth, but at a considerable distance in advance of the central region they become weaker and more and more superficial. In the rear of a storm area the westerly winds which blow may be regarded as the upper current of air extending down to the earth's surface, and accelerated by the flow of air in toward the area of low low pressure, but weakened by the retarding effect of friction.

In the trade-wind regions the easterly surface winds are the permanent and normal condition of the atmosphere, and equally so are the southwest winds that prevail above. The line that divides the north tropical from the north temperate climate (the meteorological Tropic of Cancer) moves northward during the summer months, so that, for instance, the ocean in the latitude of New Orleans, during the summer, is generally covered by easterly winds, while high above the southwest wind prevails. This normal movement north in summer and south in winter of the system of winds is, however, so greatly modified by the effect of continental heat upon the distribution of the atmosphere, and consequently upon the winds, that it would be not improper to say that in winter the winds tend to circulate around the center of continents in the direction of the motion of the hands of a watch, in the summer in the contrary direction. The disturbing effect of the unsymmetrical distribution of continents and oceans thus transforms into mingled vertical and horizontal currents the system of three simple vertical currents that Ferrel has shown would exist in the Temperate Zones if the earth's surface was all land or all water, of uniform absorbing and radiating power.

THE DIRECTION AND PROGRESS OF STORMS.

The previous considerations have been confined to the study of a few isolated charts of the weather. But in comparing the indications of a series of these, constructed as are those of this office, thrice daily, the student is at once struck with the regularity with which the areas of stormy and clear weather move over the surface. The lines of high and low pressure, the areas of high and low temperature, &c., are in continual motion generally to the eastward, except for the regions south of 30° of latitude, where the movement is westward in summer; they may change in their details, but their features are always identifiable in each successive chart until they have passed the limits of the map and other phenomena have succeeded. The rapidity of the easterly movement may occasionally amount to fifty miles per hour, but probably averages less than thirty miles.

In this connection it may be well to notice an important generalization recently published by the London Meteorological Office, as deduced from the logs and special observation of the Cunard steamships, plying between New York and Liverpool, i. e., that a vessel bound to the westward meets advancing areas of low pressure, and the observer finds that his barometer falls and rises again more rapidly than it would were he on shore, while an observer on board of a ship bound to the eastward has just the reverse experience. Instances occur in which steamers bound from New York to Liverpool overtake severe cyclones, and sometimes outstrip the eastward moving area of low pressure—a fact which, taken with what has been now advanced, clearly indicates that a number of successive barometrical depressions, each with its own cyclonic wind-system, are moving across the Atlantic, somewhat after the manner in which eddies pursue each other down the current of a river. The movement of these areas of low pressure, for both hemispheres, has been observed, on the ocean, to be eastward between the latitudes of 35° and 50°.

The controlling influences which determine the course of a given storm-center, or, in fact, any area, whether of clear, or warm, or cloudy weather, may be considered as follows:

1. There is a decided tendency of areas of low pressure to move northward more rapidly than southward, and the reverse for areas of high barometer. These tendencies are, respectively, strongest in the latitude of 45°. This principle, which is a deduction from the mathematical theory of the atmospheric currents, is confirmed by observation.

2. Storms of considerable extent disturb the atmosphere to a sufficient height to have their course determined by that of the upper currents of air; i. e., the south-west current in the North Temperate Zone.

3. Storms of less extent, for instance the local summer thunder-storms, are carried along by the general winds of the lower strata of air; these, however, are determined by the existence of the continental and oceanic areas of high and low pressure, whose changes from month to month may be seen in the charts of monthly isobarometric lines. Thus it is that, with but very few exceptions, the storms that have been traced to any distance from April to October are found to move about the tropical areas of high barometer, in the Atlantic and Pacific Oceans, in the direction corresponding to the movement of the hands of a watch, and in the contrary direction about the area of low barometer in the interior of North America; those traced during the winter months move about the area of high barometer, in the interior of the continent, in the direction of the movement of the watch-hands.

These great areas of high pressure are, however, ever varying in outline and position, thus giving rise to changes in the storm-paths.

4. The central low pressure produces a fall in the barometer, in all directions about it; wherever that fall is accompanied with a deposition of vapor, a further fall will be thereby induced; consequently, the storm-center will be drawn in that direction.

On becoming familiar with the extent of the changes that may be produced by the heat of day and the cold of night, it is learned that from September to May such daily changes rarely or never interrupt the progress of storms when they have once set in; they exert but a subordinate influence compared to that exerted over the atmosphere by the central area of low pressure, which appears to maintain the storm so long as it is supplied with moist air.

Storms may increase and decrease in severity as they move along through moist or dry regions of air, respectively; and the change is accompanied by a corresponding fall or rise in the barometer at the central area, and an expansion or contraction of the storm area.

It is especially necessary, in determining the storm path, to bear in mind the motion of the upper stratum of air. Over the country north of the Gulf of Mexico there generally prevail, (as upper currents,) in the higher regions of the atmosphere, southwest, west, northwest, or north, and, rarely, northeast winds, according to the season and the distribution of atmospheric pressure; and these are those that principally determine the general movement of storms, &c. The thin surface stratum of air is comparatively quiescent, or is, at different points, moving simultaneously in opposite directions, while overhead the whole body of air is moving onward with a far more uniform direction and velocity.

In this connection the reports of winds from the top of Mount Washington are of interest for New England, although that station can hardly be considered as always in the upper current. The phenomena of storms in the United States may be well comprehended, if we consider the office of the lower strata (within two miles) to be to carry vapor hither and thither, that of the upper strata, on the other hand, by successive aerial waves and tides, to subject the lower strata to alternate compression and dilation, thus giving rise, respectively, to clear and cloudy weather, which latter becomes at times rainy, and is then the center of a local low barometer and its system of circulating winds.

STORMS AND CYCLONES.

Whether of snow, rain, or wind, whether of greater or less violence, storms and cyclones have much similarity in their general features and behavior. Strong contrasts of temperature and of pressure, in contiguous currents of warm and cold air, mark the progress and also the origin of a storm. The Gulf Stream and the adjacent areas of colder water; the land bordering on oceans or lakes, whether frozen or open; mountains and plains, and river valleys, are examples of regions over which moist and dry or warm, and cold strata come in contact. But even more important, though imperfectly understood, are the sudden changes that take place overhead, which are apparently due to the elevation of moisture into the higher regions of the atmosphere. The storms that visit the United States may be described as of four types, as follows:

1st. The West India cyclones, originating in the southern regions of the zone of easterly trade winds, and generally east of the Windward Islands, possibly even in the Meteorological Torrid Zone, or equatorial belt of calms and rains. A very low pressure and large humidity mark their central region. Toward this the winds blow from all points, and, deflecting to the right, pursue their spiral course inward and upward; at least, this is the only satisfactory explanation that has yet been offered for the various phenomena. The moisture brought by this wind condenses as the pressure is reduced, and clouds are formed, with heavy rain.

Around the center of a cyclone an upward current is supposed to exist, and high above are formed the cirrus clouds, which stream far away in advance on the upper currents of air. These storms are carried to the north and west until they pass into the Meteorological Temperate Zone, where the prevailing south and west winds control their motions. This generally happens on or opposite the South Atlantic coast, and as the storms then pursue a course nearly parallel with the Gulf Stream, with its attendant band of warm, moist air, they produce heavy easterly gales along our Atlantic coast, and finally are lost in the Northern Atlantic, but occasionally, doubtless, reach Iceland and the coast of Great Britain.

Instances are not wanting in which these heavy storms have passed across the Gulf to Texas, thence northward to the lakes, and northeast to Maine, widening the area of disturbances, and gradually changing into extended rains, with moderate winds, thus differing from the Atlantic storms only in that their path is further westward, and that southwest gales are produced in the Eastern States.

2d. The autumn, winter, and spring rains, which generally first announce themselves on the southwest or western plains of this country, may be regarded as disturbances originating on the northern confines of the Tropical Zone, and on the Pacific slope, (as distinct from those of the preceding class that originate in the West Indies.)

From the area of high pressure on the Pacific coast of Central and North America, a volume of moist air is forced up over the Sierra Nevadas and Rocky Mountains; its moisture is deposited, and a wave of rarefied but probably dry air is started on its northeast or eastern course. No sooner does this arrive, as a wave of low barometer, over the comparatively moist air of the Mississippi Valley, than, by relieving the surface stratum of its pressure, there at once begins the condensation of its moisture, which process, if the air is not too dry, goes on rapidly increasing.

Local currents arising in this surface stratum of air feed the central area of condensation, which soon becomes hazy, and then cloudy, until rain begins. While the general progress of the storm-center will be northeastward, yet it is evident that wherever the moistest air exists there the condensation will take place the most rapidly, there the barometer will also fall the most rapidly, and thither the storm will be strongest drawn. Such storms naturally, therefore, move very rapidly up toward the lakes, and hang tenaciously over them, and move slowly away from them. In winter their course is eastward, in the early autumn northeastward.

The temperature of the upper regions must decide whether rain or snow will attend these storms. Their advance is almost invariably heralded by an increase of temperature, due apparently to latent heat evolved by the condensation going on in the circumjacent and superior air and radiated downward to the earth, and to the increased facility with which the saturated air on the surface absorbs the heat radiated by the earth.

Following the increase of temperature the cirrus clouds are observed, which geographically precede the stratus and the rain. The lowest pressure is felt on the earth after the rain has begun to fall. Although often these storms pass over without rain, until they near the lake district or Eastern States, yet their first cause may be traced back to the changes going on in the southern and western limits of the United States; at least two such have been actually followed during the four or six days occupied in passing from the California coast to Nova Scotia, and many instances are recorded of those that have passed from Texas over Lakes Superior and Huron.

The strong winds of autumn on the lakes and in the Northwestern States with little or no rain are due to disturbances that move southeastward, and whose centers are generally in British America.

3d. Well-defined, though generally weak disturbances, have been observed to pass from the north to the south, or the northwest to the southeast, but these are probably rare in the United States and probably occur only in midwinter, when the northeast winds and high pressure in British America are exceptionally strong. Continuous snow, succeeded by cold, dry weather, characterize these storms, and such a one, on one occasion during the past winter, after striking the coast of Alabama and turning eastward, ascended the Gulf Stream to the northeastward, thus coursing around the area of high pressure that had then pushed southward over the lake region.

4th. The storms which are generally confined within the United States are the northers, tornadoes, and thunder-storms. The latter are generally spread over a very narrow space, so that they may at times pass between the stations from which our reports are received. These storms evidently originate in the lower cloud stratum in local but intense differences of temperature, moisture, and pressure, and are believed in general to prevail only on the western side or in the rear of areas of high pressure. The gyratory movements of these small storms depend upon local currents and resistances rather than on the earth's rotation; they *may*, therefore, gyrate either toward the right or the left. In these storms the cumulus clouds are particularly remarkable for their height and the cirrus clouds for their small extent. The presence of a surface area of dry air is oftentimes sufficient to dissipate these storms, or to cause them to retire into the cloud regions. Similar storms form over mountain tops, and are experienced by balloon voyagers when the air is quite undisturbed below. Several such smaller storms frequently simultaneously coexist, pursuing parallel paths circulating with the general winds about the continental areas of low barometer, and the area of local storms thus corresponds very nearly to what would be an area of general rain were the temperature lower over the region. The lightning which accompanies these storms is the effect of the concentration upon large drops of water of the electricity previously distributed throughout the invisible vapor; it is considered as a *result* not a cause of storms.

5th. It has been noticed that there is a tendency in the spring and summer toward an accumulation of barometric pressure over the middle and eastern Atlantic States. When this area of high barometer moves eastward the easterly winds on its south side driving on to the coast from Maryland to Massachusetts produce clouds and occasionally severe storms of small extent, which are driven north and westward until broken up among the Appalachian Mountains.

In general, areas of high barometer prepare the way for the succeeding low pressure and high winds, and have been not inaptly termed *storm breeders*.

THE PREDICTION OF STORMS.

The wind is that element which most affects the commercial interest of the country, and, in forecasting the approach of a storm, a student at present naturally gives his principal attention to this element; the prevalence of fog, rain, or snow, and the temperature of the air, may, however, be estimated in a

general way. In making use of the Weather-Bulletin and Chart
for the purposes of prediction for any region of the country
different from that in which the student is at the moment
residing, he is, of course, cut off from the use of many local
rules which would influence his judgment were he at the time
there present, and able to know by personal observation all the
minutiæ of the atmospheric condition. If deprived of these
important helps, and forced to rely exclusively upon the bare
numerical data offered by the Bulletin, he must call to his aid
all such knowledge as is offered by the preceding brief statement of the prominent meteorological principles, such local
laws as he may know to hold good for the districts in question,
and such more general laws as have been deduced by the study
of eminent meteorologists. A number of these latter will be
found in the appendix.

The local laws referred to are now being collected at the office
of the Chief Signal Officer, and will eventually be published in
such a manner as to make them practically available. At
present such general laws are prescribed as seem most applicable to this continent.

The general direction of the wind over any region is given
by the table on page 17, supposing it to be known where the
continental and oceanic areas of high and low pressure exist.
The local wind may be deduced from the general wind by due
consideration of local friction and other resistances, as well as
of the existence of secondary and smaller areas of low or high
pressure. The force of the wind at any locality depends primarily upon the rapidity with which the pressure is changing
at that point, as indicated to some extent by the crowding of
the Isobaric lines, but is greatly modified by the local topogra-
phy, as before explained on page 19. The effect of lakes and
hills in altering the winds will at some future time be published
in detail, the present data being scarcely sufficient for most of
the stations.

The prediction of an extended storm for any portion of the
country, therefore, is reduced to the *determination of the path
pursued by the central area of low pressure*, and the rapidity
with which this will extend its influence in any given direction.

It frequently happens that two or more areas of low pressure
may be defined on the Weather-Chart; when such is the case
the areas seem to influence each other as to their extent, outline, and course of progression.

The rapidity of movement is a physical question that cannot yet be solved numerically, but, in general, it is known that for the same temperature, the moister the air, or the greater its relative humidity, the greater will be the effect of a general diminution of pressure or temperature in inducing condensation and a further fall of the barometer. When smoky haze prevails, it counteracts the tendency to formation of cloud and rain.

As regards the path pursued, it has already been stated that large areas of high pressure are known to exist in the northern hemisphere, and the small areas of low barometer that constitute the nuclei of storms have been observed to move around the borders of these greater areas; these latter, however, are by no means stationary.

Fortunately, however, the general student may be spared the trouble of the numerical estimate of the resultant of a number of mechanical forces, for the clouds themselves indicate several hours in advance the probable direction of progress of these storms, since the clouds actually form heavier and earlier in those regions in which the falling pressure or temperature has the greatest influence in condensing the moisture.

The general distribution of the principal masses of cirrus and cirro-stratus clouds, combined with the distribution of the areas over which the temperature and pressure have risen or fallen with abnormal rapidity, will safely indicate, at least for the winter months, the immediate region into which the storm will pass, and occasionally even give a premonition of its breaking up into two portions, each drawn in different directions.

The more violent winds generally follow in the rear, and on the south side of the advancing area of lowest pressure; those that precede the progress of that area may often be more dangerous, however, because of the accompanying rain, fog, &c. The latter winds are preceded by the cirrus and threatening storm-clouds; the rain that accompanies or follows these generally abates, and thus gives warning of the strong clearing-up winds.

The rapidity of progression of the area of cloud and rain varies from fifteen to sixty miles in an hour, the actual velocity varying with the influence of moisture, as explained in a previous section.

The average velocity of the currents, which determine the general direction of the progress of the nucleus of the storm, varies from twenty to forty miles hourly, and rarely reaches the higher limit.

Allusion has been made in the earlier pages to the fact that the barometric observations at very high stations cannot be combined with those taken simultaneously at lower stations in drawing our Isobaric lines. This is the result, partly, of ignorance of the laws of the diminution of moisture and temperature actually prevailing at the station, which laws vary with the hour of the day, the seasons, and the state of the weather, and still more is it the result of the extraordinary changes that are continually taking place in those upper regions, and which are but very sluggishly followed by the lowest strata. The inertia of the air conspires with the friction of the winds on the land to delay the movements in the continental storms much behind the corresponding phenomena in oceanic storms.

The barometric and thermometric changes that are reported from Mount Washington, for instance, sometimes afford sure premonitions of a change in the general character of the weather, and with great frequency foreshadow the storms that pass directly over New England.

It is by increasing the mountain stations, and by adding such balloon observations as can be made, and specially by the study of the forms, changes, motions, height, and velocity of the clouds, and of the optical phenomena of the atmosphere, that meteorologists hope eventually to arrive at a full knowledge of the regions of the air, where the severe storms are propagated.

APPENDIX.

The following appendix contains a number of important generalizations and laws given in the language of the original sources from which they are extracted.

Tables are added for the calculation of the relative humidity of the atmosphere.

PROCEEDINGS OF THE AMERICAN ASSOCIATION FOR THE ADVANCEMENT OF SCIENCE.

ELEVENTH MEETING, HELD AT MONTREAL, CANADA EAST, AUGUST, 1857.

LEADING PRINCIPLES OF REDFIELD'S THEORY OF STORMS AS DEVELOPED BY HIM FROM 1831 TO 1857.

That all violent gales or hurricanes are great whirlwinds in which the wind blows in circuits around an axis either vertical or inclined; that the wind does not move in horizontal circles, as the usual form of his diagrams would seem to indicate, but rather in spirals toward the axis, a descending spiral movement externally and ascending internally.

That the *direction of revolution* is always uniform, being from right to left, or against the sun, on the north side of the equator, and from left to right, or with the sun, on the south side.

That the *velocity of rotation* increases from the margin toward the center of the storm.

That the whole body of air subjected to this spiral rotation, is at the same time *moving forward* in a path at a variable rate, but always with a velocity much less than its velocity of rotation; being at the minimum hitherto observed as low as four · miles, and at the maximum forty-three miles, but more commonly about thirty miles per hour, while the motion of rotation may be not less than from one hundred to three hundred miles per hour.

That in storms of a particular region, as the gales of the Atlantic or the typhoons of the China Seas, *great uniformity exists in regard to the path* pursued; those of the Atlantic, for example, usually issuing from the equatorial regions eastward of the West India Islands, pursuing at first a course toward the north-west as far as the latitude of 30 degrees, and then gradually wheeling to the north-east, and following a path nearly parallel to the American coast, to the east of Newfoundland, until they are lost in mid-ocean; the entire path when delineated resembling a parabolic curve whose apex is near the latitude of 30 degrees.

That their *dimensions* are sometimes very great, being not less than one thousand miles in diameter, while their path over the ocean can sometimes be traced for three thousand miles.

That the *barometer* at any given place falls with increasing rapidity as the center of the whirlwind approaches, but rises at a corresponding rate after the center has passed; and finally,

That the phenomena are more uniform in large than in small storms, and more uniform on the ocean than on the land.

TENTH MEETING, HELD IN ALBANY, NEW YORK, AUGUST, 1859.

V.—METEOROLOGY.

I.—ON THE SPIRALITY OF MOTION IN WHIRLWINDS AND TORNADOES, BY W. C. REDFIELD, OF NEW YORK.

I. An aggregated spiral movement, around a smaller axial space, constitutes the essential portion of whirlwinds and tornadoes.

II. The course of spiral rotation, whether to the right or left, is one and the same in this respect throughout the entire whirling body, so long as its integrity is preserved; but the oblique inclination which the spiral movement also has to the plane of the horizon is in the opposite directions, as regards the exterior and interior portions of the revolving mass. Thus, in the more outward portion of the whirlwind the tendency of this movement is obliquely downward, where the axis is vertical, but in the interior portion the inclination or tendency of the spiral movement is upward. This fact explains the ascensive effects which are observed in tornadoes, and in more diminutive whirlwinds.

III. Owing to the increased pressure of the circumjacent air, in approaching the earth's surface, the normal course of the gradually descending movement in a symmetric whirlwind is that of an involuted or closing spiral, while the course of the interior *ascending* movement of rotation is that of an evolved or opening spiral; hence the horizontal areas of the higher portion of the whirl exceed greatly those of its lower portions.

IV. The area of the ascending spiral movement in the vortex, as it leaves the earth's surface, is by far the smallest portion of the whirling body, for the reason that the rotation here is proportionally more active and intense, being impelled by the aggregated pressure and movement of the more outward portion of the whirlwind as it converges from its larger areas on all sides, by increasingly rapid motion, into the smaller areas of ascending rotation.*

That this interior portion of the whirl resembles an inverted hollow cone or column, with quiescent and more rarefied air at its absolute center, may be inferred from the observations which have been made in the axial portions of the great' cyclones. Into this axial area of the tornado, the bodies forced upward by the vortex cannot fall, but are discharged outwardly from the ascending whirl. The columnar profile of this axial area sometimes becomes visible, as in the water-spouts, so called.

V. Accessions caused by circumjacent contact and pressure are constantly accruing to the whirling body, so long as its rotative energy is maintained. A correlative diffusion from its ascending portion must necessarily take place toward its upper horizon, and this is often manifested by the great extent or accumulation of cloud, which results in this manner from the action of the tornado; in other words, there is a constant discharge from the whirling body in the direction of least resistance.

VI. The spirality of the rotation and its inclination to the horizon, in the great portion of the whirl, which is exterior to its ascending area, are not ordinarily subject to direct observation; nor is the outline or body of the more outward portion of the whirlwind at all visible otherwise than in its effects.

* The law of increment in the velocity of the whirlwind, as it gradually converges into lesser areas, by its spiral involution, is that which pertains to all bodies when revolving around inferior foci, toward which they are being gradually drawn or pressed nearer and nearer in their involute course, the line of focal or centripetal pressure thus sweeping *equal areas in equal times*, at whatever diminution of distance from the center, except as the velocity may be affected in degree by the resistance of other bodies.

Such resistance is of little effect in a tornado, because its revolving mass is mainly above all ordinary obstacles, such as orchards and forests, into which the spiral *descending* and accelerated blast, near the contracted extremity of the inverted and truncated cone of the whirl, penetrates with constant freshness and intensity of force, already acquired in the higher and unobstructed region.

VII. In aqueous vortices the axial spiralities of the exterior and interior portions of the whirl are in reverse direction to those in the atmosphere, the descending spiral being nearest to the axis of the vortex. Hence lighter bodies, and even bubbles of air, are often forced downward in the water, in the manner in which heavier bodies are forced upward in the atmosphere.

FOURTH METEOROLOGICAL REPORT, BY PROFESSOR J. P. ESPY, WASHINGTON, JULY, 1854.

VIII.—GENERALIZATIONS.

1. The rain and snow storms, and even the moderate rains and snows, travel from the west toward the east, in the United States, during the months of November, December, January, February, and March, which are the only months to which these generalizations apply.

2. The storms are accompanied with a depression of the barometer near the central line of the storm, and a rise of the barometer in the front and rear.

3. This central line of minimum pressure is generally of great length from north to south, and moves side foremost toward the east.

4. This line is sometimes nearly straight, but generally curved, and most frequently with its convex side toward the east.

5. The velocity of this line is such that it travels from the Mississippi to the Connecticut river in about twenty-four hours, and from the Connecticut to St. John, Newfoundland, in nearly the same time, or about thirty-six miles an hour.

6. When the barometer falls suddenly in the western part of New England, it rises at the same time in the valley of the Mississippi, and also at St. John, Newfoundland.

7. In great storms the wind for several hundred miles on both sides of the line of minimum pressure blows toward that line directly or obliquely.

8. The force of the wind is in proportion to the suddenness and greatness of the depression of the barometer.

9. In all great and sudden depressions of the barometer there is much rain or snow; and in all sudden great rains or snows

there is a great depression of the barometer near the center of the storm, and rise beyond its borders.

10. Many storms are of great and unknown length from north to south, reaching beyond our observers on the Gulf of Mexico and on the northern lakes, while their east and west diameter is comparatively small. The storms therefore move side foremost.

11. Most storms commence in the "far west," beyond our most western observers, but some commence in the United States.

12. When a storm commences in the United States the line of minimum pressure does not come from the "far west," but commences with the storm, and travels with it toward the eastward.

13. There is generally a lull of wind at the line of minimum pressure, and sometimes a calm.

14. When this line of minimum pressure passes an observer toward the east, the wind generally soon changes to the west, and the barometer begins to rise.

15. There is generally but little wind near the line of maximum pressure, and on each side of that line the winds are irregular, but tend outward from that line.

16. The fluctuations of the barometer are generally greater in the northern than in the southern parts of the United States.

17. The fluctuations of the barometer are generally greater in the eastern than in the western part of the United States.

18. In the northern parts of the United States the wind generally in great storms sets in from the north of east and terminates from the north of west.

19. In the southern parts of the United States the wind generally sets in from the south of east and terminates from the south of west.

20. During the passage of storms the wind generally changes from the eastward to the westward by the south, especially in the southern parts of the United States.

21. The northern part of the storm generally travels more rapidly toward the east than the southern part.

22. During the high barometer on the day preceding the storm it is generally clear and mild in temperature, especially if very cold weather preceded.

23. The temperature generally falls suddenly on the passage of the center of great storms, so that sometimes, when a storm

is in the middle of the United States, the lowest temperature of the month will be in the west on the same day that the highest temperature is in the east.

Some of the storms, it is true, are contained entirely, for a time, within the bounds of my observers, and in that case the minimum barometer does not exhibit itself in a line of great length, extending from north to south, but it is confined to a region near the center of the storm, and travels with that center toward the eastward.

From these experiments it may safely be inferred, contrary to the general belief of scientific men, that *vapor permeates the air from a high to a low dew-point with extreme slowness*, if, indeed, it *permeates it at all;* and in meteorology, it will hereafter be known that *vapor rises into the regions where clouds are formed only by being carried up by ascending currents of air containing it.*

EXTRACT FROM LETTER OF S. C. WALKER, MARCH, 1837.

Dr. Franklin first ascertained that all storms travel toward the northeast.

Mr. Espy's researches led him to believe that this constancy of direction is confined to our winter storms and summer tornadoes. He has also found, from observations furnished by Professor Hamilton, that in the winter season the rise of barometer, as well as the presence of the center of the storm, takes place at Nashville, Tennessee, about twenty-four hours earlier than at Philadelphia. And here I would remark that Dr. Emerson is the first observer, so far as my knowledge extends, who noticed that a great rise of the barometer is a prelude to a north-easterly storm—a conclusion to which Mr. Espy has arrived *a priori* from his theory of storms.

This conclusion is in direct opposition to popular opinion, and, indeed, to that of most philosophers, who have marked set fair on the barometer at one inch above the mean.

ABSTRACT OF RESULTS OF PROFESSOR FERREL'S INVESTIGATIONS.

The valuable mathematical essay of Professor William Ferrel, published in 1856, (and a second edition in 1860,) " On the motions of fluids and solids on the surface of the earth," specifies the following general laws as applicable to our atmosphere:

GENERAL MOVEMENT OF THE ATMOSPHERE.

Assuming that there is absolutely no friction between the atmosphere and the face of the earth, and regarding the latter as a sphere rather than an oblate spheroid, we have, for the general condition of the atmosphere, the following conclusions:

The atmosphere, however deep it may be at the equator, cannot exist at the poles.

The exterior surface of the atmosphere would be slightly depressed at the equator, and have its maximum height about the parallel of 35 degrees, and meet the surface of the earth near the poles.

At the latitude of maximum height the atmosphere would have no motion east or west.

Between the parallel of 35 degrees and the poles the motion would be eastward, but betweeen those parallels and the equator toward the west.

LOCAL MOVEMENTS.

Under the same assumption of no friction, we have for a small circular portion of air rotating horizontally on the earth's surface:

The air, however deep it may be at the extreme boundary, cannot exist at the center.

The upper surface of the revolving portion of air will be very slightly convex, and meet the surface of the earth near the central axis of revolution.

There will be no gyratory motion at the region of maximum height of the air.

The inner part of the fluid will gyrate from right to left, (i. e., opposed to the motion of the hands of a watch,) but the external part from left to right.

If the fluid gyrate from right to left, the whole mass has a tendency to move toward the north; but if from left to right, toward the south.

In whatever direction a body moves on the surface of the earth there is a force arising from the earth's rotation which deflects it to the right in the northern hemisphere, but to the left in the southern.

4

GENERAL MOVEMENT, TAKING ACCOUNT OF FRICTION AND
VARYING DENSITY.

Although the preceding results, when applied to the atmosphere, are very much modified by the resistances of the earth's surface, yet they will be of great advantage in explaining its general motions; for as there can be no resistance until there is motion, the atmosphere must have a tendency to assume, in some measure, the same motion and figures as in the case of no resistances. Hence, toward the poles, the general motions of the atmosphere must be toward the east, and in the torrid zone toward the west; there must also be a comparatively small depression at the poles and at the equator.

There must be a region of calms about the poles, and a belt of calms at the equator, (the latter belt lying a short distance north of the exact equator.)

The belt of calms, which, in the case of no friction, it was previously shown, must exist at the parallels of 35 degrees, will be moved toward the equator nearly to the parallels of 30 degrees (and nearer the equator in the northern than in the southern hemisphere.)

The lesser friction of air moving over air than over the earth causes an additional accumulation of atmosphere at the tropical belts, the outflow of which, combined with the westerly and easterly motions of the atmosphere, gives rise to the fresh north-east trade-winds of the northern hemisphere, and to the south-west surface-currents of the temperate zone.

The cold air of the polar regions gives rise to the very superficial, weak, north-easterly winds of the arctic zone.

A belt of calms must exist within the polar circles. The system of belts of calms and winds on the earth's surface must be distorted, from absolute symmetry, by the influence of continents, and oceans, and ocean currents, and the respective belts must change their positions somewhat during the year as the sun varies his declination; moving southward in the fall, and northward during our spring. The upper currents of air are from the south-west over the entire northern hemisphere; but a middle current [or perhaps a local superficial current] from the north-east may exist in the temperate zone.

LOCAL MOVEMENTS—CYCLONES.

Whenever, by reason of local rarefaction or other cause, an upward current is established at any place, fed by moist surface-

51

currents, the surrounding atmosphere assumes a gyratory motion, but the resistances cause a calm at the exact central area, the most rapid motion being on the immediate outer limit of this area. The contrary gyrations above indicated, as existing on the outer rim of the cyclone, where no frictional resistance is present, are in this case generally destroyed by such resistance. The gyrations will be from right to left in the northern hemisphere. At the equator there will be no gyratory motion, and consequently no cyclones. The effect of the earth's rotation in determining the direction of rotation of a cyclone is very small where the disturbance extends over a small area, hence tornadoes, properly so-called, depend for their gyrations more upon the initial state of the atmosphere, and may rotate in either direction; hence tornadoes, but not cyclones, may be experienced at the equator.

The force that maintains the gyrations of a cyclone being one in constant action, these perpetuate themselves from hour to hour, and for many days; while in tornadoes the resistances of the earth and air soon overcome the initial gyratory tendency.

The great depression of the barometer in tornadoes and cyclones is caused, not so much by the rarefaction of the air by expanding moisture, though this ordinarily gives the first start to the whirlwind, but principally by the centrifugal (not the tangential) force due to the rapid motion of the particles of air near the center.

As there is less resistance in the upper strata, the rapid gyratory motion commences there first.

The interior portions of cyclones always gyrating, as they do, from right to left in the northern hemisphere, must always move toward the north pole; while between the equator and the tropical calm-belt they are carried westward by the general motion of the atmosphere, but after passing this belt the general atmospheric movement carries them eastward.

Near the equator they must move slowly toward the poles, but after passing the tropical calm-belt, the motion of progression must be accelerated.

The progression of small tornadoes is dependent almost entirely upon that of the current of air in which they exist, and the general tendency of all small disturbances is to run into the larger belts of low pressure.

RESULTS OF PROFESSOR COFFIN'S INVESTIGATIONS.

The following generalizations are deduced by Professor J. H. Coffin, in his exhaustive memoir "On the Winds of the Northern Hemisphere:"

In the arctic regions of North America, lying within the polar circle, the mean direction of the wind is about north-north-west, and well defined.

Between the parallels of 60 degrees and 66 degrees there appears to be a belt of easterly or north-easterly winds, whose pole is at about latitude 84 degrees and longitude 28 degrees west of Washington.

Passing south of this circle we find a belt of westerly winds, about 23½ degrees in breadth, entirely encircling the globe, and the poles of whose southern and northern limits very nearly coincide with that of the preceding belt.

Near the limits which divide this zone from its neighbors the progressive motion of the wind is very small; the progression is less in Europe than in America.

Passing south of this [the temperate] zone, we find that contiguous to it the winds are on the whole easterly, yet quite irregular, and having a very small progressive motion.

Farther south we fall in with the well-known north-easterly trade-winds, showing more decided prevalence between latitude 10 degrees and 25 degrees than nearer the equator.

On each side of the Atlantic Ocean there is a systematic change in the winds prevailing during the different seasons of the year, similar to the monsoons of Asia.

On the Atlantic coast of North America the monsoon character is more marked than on the European coast, and more marked on the coast than in the interior; but again becomes well marked as we near the elevated plains west of the Mississippi.

BLODGET'S CLIMATOLOGY OF THE UNITED STATES.

Dr. Gibbons has noticed with great care, at San Francisco, the course of the higher strata of clouds—the cirrus and the very high stratus—when they were visible, and has found them to come uniformly from some westerly point, as he had also observed for many years at Philadelphia. The writer has long observed the same facts in western New York, where an average of not more than one instance annually occurs of clouds in the higher strata moving from any other than a westerly point. During three years of very careful registry directed to this particular point, but three instances of a contrary direction were observed; and these were during the prevalence of extensive and disastrous storms on the Atlantic coast. The lower clouds are from various points, and the wind is quite variable during the greater storms, two strata of different movement often lying beneath that from the west, yet the stratum from a westerly point usually deposits the rain, and when it ceases the rain-fall ceases, though the lower strata may continue to run on the wind twenty-four hours or more longer.

ABSTRACT OF PROFESSOR BUCHAN'S RESULTS.

The invaluable charts of Mr. Buchan, showing the average barometric pressure and winds for each month of the year, together with the many detailed generalizations, can only be referred to here. The following, however, seem to sum up his results:[*]

"An examination of the isobaric and wind charts for the months shows, as has been already pointed out, that where there is a mean low pressure, such as occurs in the north of the Atlantic in the winter months, and in the center of Asia in the summer months, thither-ward the winds tend in all directions in an in-moving spiral course; and where there occurs a mean high pressure, as in the center of Asia in winter and in the Atlantic, between Africa and the United States, in summer, out of this space the winds flow in all directions, or they appear to be thrown out from the space of high pressure in a manner exactly the reverse from that by which they are drawn inward upon a space of low pressure."

ABSTRACT OF PROFESSOR CHASE'S RESULTS.

Pliny Earle Chase, in the proceedings of the American Philosophical Society, 1871, March 3, among other results gives the following:

"The wind, especially in the southern states, often blows directly in the line of the greatest barometric gradient. But even in such cases, after a few hours' continuance, it tends toward the azimuth indicated by Buys Ballot's law.

* Transactions of The Royal Society of Edinburgh, vol. xxv, Mr. Alexander Buchan on the mean pressure of the atmosphere, &c.

"The isobaric lines are therefore often of less relative importance than the gradients in forming forecasts.

" Currents with an anti-cyclonic tendency, controlled by areas of high barometer, are notably common.

"Our recent storms have been anti-cyclonic, and there seems some reason for supposing that anti-cyclones are the usual weather-breeders, even of such of our land storms as become more or less cyclonic after they are fully developed.

"The precipitation of vapor of course gives rise to local cyclones, which, however, may be easily and speedily overborne by the grand anti-cyclonic whirls of a half million miles or more in area."

EXTRACTS FROM " BAROMETER MANUAL," COMPILED BY ROBERT H. SCOTT, FOR THE METEOROLOGICAL OFFICE, LONDON, 1871.

RULES TO EXPLAIN THE INDICATIONS OF THE INSTRUMENTS.

It should always be remembered that changes in weather generally give signs of their coming, for the instruments are affected before the wind actually begins to blow or the rain to fall; thus they may be said to enable us to feel the pulse of the atmosphere. It must not be forgotten that the length of time which passes between the first appearance of a change of weather and its actual setting in are not the same. It is much greater when a south-west wind is going to succeed a north-east wind than when the opposite change is about to take place. We shall see, a little further on, why this is the case, and also how the appearance of the sky will aid us in forming an opinion as to probable weather.

The general principles on which the following rules are founded have been laid down by Professor Dové, of Berlin, on the basis of a long series of observations, which were made at several stations situated in the north temperate zone, between the parallels 49 degrees and 65 degrees, to which regions they specially refer. The rules themselves may be shortly stated thus:

The average height of the mercury in the barometer, at sea level, in the British Islands, is about 29.9 inches. If the barometer rises steadily above its mean height while the weather gets colder and the air becomes drier, north-westerly, northerly, northeasterly winds, or less wind, less rain or snow, may generally be expected. On the contrary, if the barometer falls while the weather gets warmer, and the air becomes damper, wind and rain may be looked for from the south-east, south, or south-west.

The deviations from these general principles which are noticed correspond to the various changes of weather.

If the weather gets warmer while the barometer is high and the wind north-easterly, we may look for a shift of wind to the south. On the other hand, the weather sometimes becomes colder while the wind is south-westerly and the barometer low, and then we may look for a sudden squall, or perhaps a storm, from the north-west, with a fall of snow, if it be winter time.

No absolute laws for weather can, however, be laid down; the most striking exceptions to the rules are those noticed by Admiral Fitz Roy. They occur with north-east winds, which sometimes bring rain, or sleet, or snow, especially during gales, although the barometer may be high and rising. On the other hand, when the wind is north-easterly and light, and the barometer begins to fall, rain may set in before the wind changes to east or east-south-east.

Besides these rules for the instruments, there is a rule about the way in which the wind changes, which is very important. It is well known to every sailor, and is contained in the following couplet:

> " When the wind shifts against the sun,
> Trust it not, for back it will run."

The wind usually shifts *with the sun*, i. e., from left to right* in the northern hemisphere. A change in this direction is called *veering*.

Thus an east wind shifts to west through south-east, south, and south-west, and a west wind shifts to ough north-west, north, and north-east. If the wind shifts the opposite way, viz., from west to south-west, south, and south-east, the change is called backing, and it seldom occurs, unless when the weather is unsettled.

However, slight changes of wind do not follow this rule exactly; for instance, the wind often shifts from south-west to south and back again.

In most parts of the world it has been observed that there are two prevailing wind-currents, which vary with the circumstances of the place, but are, on the whole, nearly opposite each other.

---

* In the southern hemisphere motion *with the sun* is, of course, from right to left.

In these islands these directions are about north-east and south-west, and the latter of these winds blows for about ten times as many days in the year as the other does.

What is it that causes these winds to blow and makes them so different from each other, as we know them to be? The simplest account of them is that the air is always flowing toward the equator from the poles, and back again. It then forms two great currents: one is called the polar current, as it flows from the direction of the pole, and is felt here as a north-east wind; the other is called the equatorial current, as it flows from the direction of the equator, and is felt here as a south-west wind.

The air of the polar current has been chilled, and is heavy, cold, and dry; while it is blowing the barometer is high and the weather usually dry.

The air of the equatorial current has been heated, and is light, warm, and moist; while it is blowing the barometer is low and the weather usually wet.

If we keep the idea of these two great wind-currents clearly in our heads, we shall easily understand most of the signs of the weather which are noticed.

The air of the equatorial current is lighter than that of the polar, and so southerly winds will begin to blow aloft before they are felt on the ground, while northerly winds will begin to blow close to the ground. Accordingly south-west winds give much more warning of their coming than north-easterly ones.

The south-west wind will often show itself first by long streaks of cirrus clouds at a great height, called "mare's tails;" or, when a gale is very near, by driving scud.

Signs of weather, such as those just noticed, are important to any one watching for changes, as they will enable him to confirm or modify the opinions formed from the behavior of his instruments. As to the instruments themselves, we have already seen that when the barometer rises, owing to a change of wind, the weather usually becomes colder; while when the barometer falls, owing to a change of wind, the weather usually becomes warmer. If the barometer be high, (above 30.5,) and *remain steady* for some days, it is because there is, so to speak, a surplus of air at the place. The wind will be light, and the weather will probably be dry. A gale can set in only when the air flows away, and it will not at first be severe at the place. If the barometer be low, (below 29.0 inches,) and remain steady, there is a deficiency of air at the place. The wind will be light, also, but the weather will probably be cloudy and wet. However, there may be fine weather for a short time, what is called a "pet day," but there is great danger of a serious storm, because the air will try to force its way into the district where the readings are low, and increase the pressure there so as to restore the atmospherical equilibrium.

If the barometer rises slowly from a low level, the weather may become drier, and the wind lighter, or perhaps die away. There may also be local fogs.

If the barometer falls gradually from a high level, the weather may become wetter and more unpleasant, and there will never be a certainty of having a fine day, though there need not be much wind.

In general, whenever the level of the mercury continues steady, we may expect settled weather, but when it is unsteady we must look for a change, and perhaps a serious gale. A sudden rise of the barometer is very nearly as bad a sign as a sudden fall, because it shows that atmospherical equilibrium is unsteady. In an ordinary gale the wind often blows hardest when the barometer is just beginning to rise, directly after having been very low.

It must never be forgotten that it is impossible for any one to interpret the meaning of all the changes in his barometer, at first, or perhaps for a day or two, inasmuch as he requires to learn what is going on at stations in his neighborhood, for without this information he cannot know whether these changes are due to mere local causes, or are the first symptoms of the approach of a more serious disturbance. A storm may be raging at a comparatively short distance from him, but his barometer, *taken by himself*, will not necessarily enable him to detect its existence.

However, in many cases, a good guess at what is likely to happen may be formed by an experienced observer who watches his instrument closely, records its indications on such a form as shown at Plate IV,* and interprets them by the rules provided in this manual. He will, however, require to call to his aid not only observations of the temperature and dampness of the air, but all his experience as to the influence of the several seasons, the ordinary character of the water at the place, and the local signs of its change.

The Daily Weather Reports issued by the Meteorological Office are calculated to render important service to any one who

---

* See "Barometer Manual," London, 1871.

wishes to study weather; they contain observations made daily at 8 A. M., at twenty British and about as many foreign stations. Great care has been taken to insure the accuracy of these reports, and the result is that a great mass of information of very great value is published every day.

The table shows the readings of the barometer and dry and wet-bulb thermometers, the direction and force of the wind, &c., and from it a very good idea may be gathered of the weather which is actually prevailing on or near our coast.

As regards the use which may be made of these reports, a most important principle has been discovered of late years.

Professor Buys Ballot, of Utrecht, and others have shown that we can tell with considerable certainty what wind may be expected to blow at any place if we know the readings of the barometer, taken a short time previously, at a number of stations situated within a distance of, say, one hundred or two hundred miles from that place.

The rule is: Stand with your left hand toward the place where the barometrical reading is lowest, and your right hand toward that where it is highest, and you will have your back to the direction of the wind which will blow during the day.

Thus the wind may be expected to be—

Easterly $\left\{\begin{array}{l}\text{when the pressure}\\\text{is highest in the}\end{array}\right\}$ north $\left\{\begin{array}{l}\text{or lowest}\\\text{in the}\end{array}\right\}$ south.

Southerly........do..........east ......do.....west.

Westerly ....do..........south.....do.....north.

Northerly........do..........west......do.....east.

The force of the wind on each day bears some proportion to the amount of difference in barometrical readings noticed between any two stations situated near the place where the wind was felt. Thus we find that it has been shown that a westerly gale hardly ever blows in the British Isles, unless, at least a few hours before, the pressure in the north of Scotland is half an inch less in amount than it is on the south coast of England.

We shall return to this subject when dealing with weather telegraphy. At present it is sufficient for us to say, with reference to the principles above laid down for the behavior of the instrument, that whenever a storm is blowing, the level of the barometer will be very different at stations near each other, so that as the storm travels across the country the barometer at any station will show signs of its coming and going by the mercury sinking or rising in the tube. This shows us why it is when the barometer is steady there is no great likelihood of a sudden change of weather, while when it is changing quickly, there is great danger of the wind freshening to a gale.

WEATHER TELEGRAPHY.

The facilites afforded us by means of telegraphy, for comparing observations taken simultaneously at several stations, have revealed to us great differences, even between adjacent stations, as regards the instrumental readings, and the actual phenomena observed under the various conditions of weather. In seeking to assign causes for these differences, we have been greatly assisted by applying the principle to which allusion has already been made under the name of Buys Ballot's law.

The immediate result of the law is to show that whenever barometrical readings are lower over any area than over those adjacent to it, the air will sweep round that area as a center, and the direction of its motion will be opposite to that of the hands of a watch. Conversely, the air will sweep round an area of relatively high barometrical readings in the direction in which the hands of a watch move. The former of these motions are said to be *cyclonic*, the latter *anti-cyclonic*. These words are derived from the word "cyclone," the general name for hurricanes and typhoons, in all which storms the motion of the air takes place around an area of diminished barometrical pressure.

We see, therefore, that the existence of a deficiency of atmospheric pressure, or what is termed a barometrical depression, over any district, is accompanied by a cyclonic movement in the air in the neighboring districts.

The actual movement of the air has no reference, either in direction or velocity, to the absolute readings of the barometer, at the point where it is lowest, or to the distance of the particles of air which are in motion from that point, but is related almost entirely to the distribution of pressure in accordance with Buys Ballot's law. The law gives the direction of motion, and its truth for these islands* and the adjacent parts of the earth's surface is incontestable; it appears, moreover, to hold good generally.

The velocity of the air depends, at least in a great measure, though not absolutely, on the difference of barometrical readings over a given distance, or on what is termed the barometrical "gradient."

The gradients adopted by the meteorological office are ex-

---

*British islands.

pressed in hundredths of an inch of mercury per fifty geographical miles.

To apply the same principles to the weather of the British Islands generally, it may safely be asserted that no storm of any serious extent is ever felt over the United Kingdom, unless there be an absolute difference in barometrical readings, exceeding half an inch of mercury between two of our stations.*

The difference in readings between Rochefort and Aberdeen, on the 1st of February, 1868, when a tremendous westerly gale was blowing, was as much as 1.76 inches; the reading at Rochefort being 30.16, and that at Aberdeen 28.40 inches. These figures give a gradient of 13.5 over the entire distance of six hundred and seventy-three miles, and we find that gales were reported from sixteen stations that morning. If these simultaneous barometrical readings for any considerable tract of country, such as that represented in the daily weather reports, which embrace these islands† and the adjacent coasts of the continent, be entered on a chart, a simple inspection of that chart is sufficient to show the direction and probable force of the winds felt over the entire district. Each of these elements will, however, be modified to a certain extent by the irregularities of the surface, for we can only feel the currents of the lowest stratum of the atmosphere, which are liable to be checked and deflected by mountains, and, in their passage over plains, to be seriously retarded even by the woods with which those plains are covered.   *   *   *   *

All the storms which we feel are accompanied by a considerable relative reduction of pressure; and these barometrical depressions travel over the country, carrying their own wind-system with them. If, therefore, we could determine beforehand the direction of advance, and the rate of motion of each successive area of depression as well as its shape, its gradients in each direction, and the rate of their increase or decrease in intensity, we should have made a considerable advance toward forecasting weather. Of these several conditions our knowledge is very incomplete. The attempts which have hitherto been made to lay down laws for any of these undetermined quantities have met with a very limited amount of success. The *shape* of the

---

* Local storms, which occasionally do great damage, may be felt when the barometrical disturbance is itself only local, and when the actual amount of difference between the extreme readings is less than half an inch, although the gradients for a short distance may be high.

† British islands.

area of depression is far from uniform, and is liable to modification, first, according to the character of the ground over which it passes; and, secondly, according to the conditions of pressure in the neighborhood.

The *direction of advance* takes place most usually from some point between south-west and north-west, but not infrequently lies in a different direction, and it is stated that occasionally a motion even from the eastward has been recognized. The velocity of motion varies from five or six miles an hour to as much as sixty or seventy; this latter rate of motion having been reached by the storm of December 16, 1869. * *

Owing to the extreme sensitiveness of the thermometer to changes of weather, it has been frequently proposed to consider its indications as fully equal in importance as those of the barometer; but great caution is necessary in acting on this idea. The accuracy of thermometrical observations depends upon a great many conditions, such as aspect, exposure to the air, elevation above sea-level and above the surface of the ground, all of which are immaterial or can be allowed for in dealing with the barometer.

THE "POLAR-BANDS" AS STORM INDICATORS, BY DR. M. A. F. PRESTEL.

A. Von Humboldt is the first to have directed attention to the uniformly broken groups of delicate clouds (cirro-cumulus) and cloud striæ, (cirro-stratus,) and to describe them under the name of polar-bands, (bandes polaire,) because their perspective points of convergence (vanishing points) very frequently lie in the magnetic poles, so that the parallel series of little clouds and bands follow the magnetic meridian.

The rays of the auroræ polaris show similar vanishing points, and not unfrequently one subsequently finds on the extinction of the polar lights the cirrus bands in the direction of these rays; this indication has called out most varied associations of ideas as well as words, since polarity, pole, pole of the winds, pole of the cold, &c., do not fail by many persons to excite the lively activity of the imagination.

The peculiarity of this enigmatical phenomenon, as A. Von Humboldt very appropriately specifies, is the variability, or at other times the regular change of the vanishing points. Ordinarily the bands are perfectly formed only in one direction, and as they move one sees them directed first from south to north, then gradually from east to west. They arise at times of great clearness of the sky. According to Humboldt, they are much more frequent under the tropics than in the temperate and cold zones. The observation that the original direction of the polar-bands from south to north gradually changes to that from east to west, applies also to the southern hemisphere.

Over north-west Germany polar-bands extend from south to north, or also from south-south-east to north-north-west, then change slowly into the position from south-south-west to north-north-east, and not unfrequently the vanishing points move still further to the east and west points. To the author of the Cosmos no relation seemed likely between the polar-bands and the upper currents of air. He says: "One cannot ascribe the progress (of these vanishing points) to the change in the current in the highest regions of the atmosphere."

When such polar-bands appear over Europe the telegraphic weather reports at present make it possible to compare the simultaneous condition of the atmosphere over the whole of Europe. By this comparison I have found that an area of storm, even if still far distant, is simultaneously present in all cases where decided polar-bands and their vanishing points in the horizon are shown. The polar-bands are present then on the extreme limits of the storm area, and have a tangential direction to its limiting line. While the weather is still quiet and beautiful in the lower regions of the atmospheric ocean, the polar-bands already indicate the currents of air in the higher strata. The gradual change to eastward of the vanishing points of the bands pointing north and south is the consequence of the progress of the center of the storm. When the latter moves from the west over the Atlantic Ocean (toward Holland) the polar-bands have the direction from south to north at the first approach of the storm-area; if the center of the storm and the bands themselves advance toward the north-east, then also the apparent direction of the latter alters in a corresponding manner, and since they are at right angles to a line drawn toward the storm-center, they show to the observer the direction in which such a storm-area lies, though in many cases five hundred to a thousand miles distant, as well as that in which it progresses. If the storm-area does not pass to one side of the observer, but the center approaches him more or less directly, then always twenty-four to thirty-six hours elapse before the arrival of the storm. The storms thus announce themselves telegraphically by means of the polar bands. The remaining uncertainty with respect to their paths which may still prevail is removed in the customary manner by the barometric observations. The polar-bands should not be confounded with the fine cirrus filaments when these appear isolated. The polar bands form on the celestial hemisphere a configuration similar to that of the marks and stripes on the surface of a melon, and they always converge toward two opposite points in the horizon.—*From " Der Sturmwarner," Emden,* 1870.

TABLES.

The explanation of the accompanying tables is as follows:

Tables No. I and No. II are used for finding the moisture or relative humidity of the atmosphere at any station, when simply the indications of the hygrometer are to be had.

The hygrometer in use by the Signal Office consists of a wet and a dry-bulb thermometer.* Having noticed the height of the mercury in each of these two thermometers, the observer enters the table with the difference of the two readings (on the horizontal line at the top) and with the temperature of the wet-bulb thermometer, (denoted by *t'*.) Let him then seek first the column at the head of which stands the difference of the thermometers; then go down as far as the horizontal line, at the beginning of which stands the temperature of the wet-bulb thermometer; and the figures he finds on that line will express the relative humidity, for which he is searching.

---

*The best form of hygrometer consists of two mercurial thermometers, which, placed side by side, indicate the same temperature. When fixed together on a frame they are called the *wet and dry bulb thermometers.* The dry-bulb is a common thermometer, intended to show the temperature of the air. The wet-bulb is also a common thermometer, having its bulb covered with a piece of muslin, from which pass a few threads of darning-cotton into a small vessel containing rain-water. The water rises by capillary attraction, thus keeping the muslin constantly wet. When the air is dry, evaporation goes on rapidly from the muslin, and on account of the heat lost by evaporation, the wet-bulb indicates a lower temperature than the dry-bulb. But when the air is damp, evaporation is slower, and the difference between the two thermometers becomes smaller. When the air is completely *saturated* with moisture, evaporation ceases entirely, and the two thermometers show the same temperature. The instrument is shown in the following cut:

When the temperature in both thermometers is the same (or the difference 0°) evaporation from the moist muslin on the wet-bulb has entirely ceased, the quicksilver in its tube has ceased to fall, and, of course, the relative humidity is that of full saturation, or 100, as is shown by the second vertical column in Table No. 1.

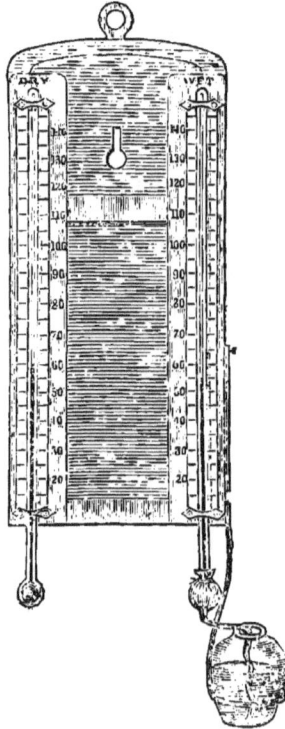

THE WET AND DRY BULB THERMOMETER.

Table No. 1 is used but seldom, because it is adapted for extreme temperatures below the freezing-point of fresh water, (32° Fahrenheit.) It gives the relative humidity in the vertical columns for differences of each half degree between the two thermometers, and when the wet bulb falls as low down as 0°. Whenever the wet-bulb stands at any figure between 0° and 9° above 0° (inclusive) use lower part of Table No. 1. In other cases table No. 2.

These tables are compiled from "Meteorological and Physical Tables, prepared for the Smithsonian Institution, by A. Guyot," third edition, Washington, 1859.

No. 1.—BELOW FREEZING-POINT; BULB COVERED WITH A FILM OF ICE.

Temperature Fahrenheit.—Relative Humidity in hundredths.

Wet-bulb thermometer t' Fahrenheit.	0°.0 Relative humidity.	0°.5 Relative humidity.	1°.0 Relative humidity.	1°.5 Relative humidity.	2°.0 Relative humidity.	2°.5 Relative humidity.
—31	100	36.0				
—30	100	39.6				
—29	100	42.9				
—28	100	46.1				
—27	100	49.0				
—26	100	51.8				
—25	100	54 4				
—24	100	56.8				
—23	100	59.0				
—22	100	61.0				
—21	100	62.6	26.9			
—20	100	64.2	30.3			
—19	100	65.9	33.5			
—18	100	67.5	36.6			
—17	100	69.0	39.5			
—16	100	70.4	42.3			
—15	100	71.8	44.9	19.4		
—14	100	73.0	47.4	23.0		
—13	100	74.3	49.8	26.4		
—12	100	75.4	51.9	29.5		
—11	100	76.5	53.9	32.5		
—10	100	77.5	55.7	35.3	15.6	
— 9	100	78.5	55.7	38.3	19.1	
— 8	100	79.4	59.4	40.6	22.5	
— 7	100	80.3	61.1	43.0	25.7	
— 6	100	81.1	62.7	45.4	28.4	12.9
— 5	100	81.8	64.5	47.6	31.7	16.4
— 4	100	82.5	65.8	40.8	34.5	19.8
— 3	100	83.2	67.1	51.7	36.9	22.8
— 2	100	83.9	68.3	53.5	39.3	25.8
— 1	100	84.5	69.5	55.3	41.6	28.6
— 0	100	85.0	71.0	57.0	43.8	31.3
0	100	85.0	70.7	57.0	43.8	31.3
1	100	85.6	71.8	58.6	46.0	33.9
2	100	86.2	73.0	60.2	48.0	36.4
3	100	86.7	74.0	61.8	50.0	38.8
4	100	87.2	75.0	63.3	52.0	41.2
5	100	87.7	76.0	64.7	53.8	43.4
6	100	88.2	76.9	66.0	55.3	45.2
7	100	88.6	77.7	67.1	56.8	47.0
8	100	89.0	78.4	68.2	58.2	48.8
9	100	89.4	79.1	69.2	59.6	50.5

5

No. 2.—RELATIVE HUMIDITY.

Difference between wet and dry bulb thermometers.

WET-BULB THERMOMETER.	1°	2	3	4	5	6	7	8	9	10	11	12	13	14	15	16	17	18	19	20	21	22	23	24	25
10	79	61	44	28	13																				
15	83	67	52	39	27	15	5																		
20	85	72	60	48	38	28	19	11																	
25	88	77	67	59	48	41	32	26	18	10															
30	89	79	69	61	52	45	37	30	26	18	11	6													
35	90	80	71	63	56	48	42	35	30	24	19	14	10	6											
40	91	83	75	67	60	54	48	42	37	32	27	23	19	14	12										
45	92	85	77	71	64	59	53	48	43	38	34	30	26	23	19	16	13	10							
50	93	86	79	73	68	62	57	53	48	44	40	36	33	29	26	23	20	18	15	13	11				
55	93	87	81	76	70	66	61	57	52	48	45	41	38	35	32	29	26	24	20	19	17	15	13	11	9
60	94	88	83	78	73	68	64	60	56	52	49	45	42	39	36	34	31	29	26	24	22	20	18	17	15
65	94	90	84	79	75	71	67	63	59	56	52	49	46	43	40	38	35	33	31	29	27	25	23	21	20
70	95	90	85	81	76	72	69	65	61	58	55	52	49	46	44	41	40	37	35	32	31	29	27	26	24
75	95	90	86	82	78	74	70	67	64	61	58	55	52	49	47	45	42	40	38	36	34	32	30	29	27
80	95	91	87	83	79	75	72	69	66	63	60	57	54	52	49	47	45	43	41	39	37	35	34	32	30
85	96	91	88	84	80	76	73	70	67	64	62	59	56	54	52	50	47	45	43	41	40	38	36	35	33
90	96	92	88	85	81	78	75	72	69	66	63	61	58	56	54	52	49	47	46	44	42	40	39	37	36
95	96	92	89	85	82	79	76	73	70	68	65	62	60	58	56	53	51	49	48	46	44	42	41	39	38
100	96	93	89	86	83	80	77	74	71	69	66	64	62	59	57	55	53	51	50	48	46	44	43	41	40

THE CAUTIONARY SIGNAL.

The Cautionary Signal of the Signal Service, U. S. Army—a red flag with black square in the center by day and a red light by night—displayed at the office of the observer, and other prominent places throughout any city, signifies:

1. *That from the information had at the Central Office in Washington, a probability of stormy or dangerous weather has been deduced for the port or place at which the cautionary signal is displayed, or in that vicinity.*

2. *That the danger appears to be so great as to demand precaution on the part of navigators and others interested—such as an examination of vessels or other structures to be endangered by a storm—the inspection of crews, rigging, etc., and general preparation for rough weather.*

3. *It calls for frequent examination of local barometers, and other instruments, by ship captains, or others interested, and the study of local signs of the weather, as clouds, etc., etc. By this means those who are expert may often be confirmed as to the need of the precaution, to which the Cautionary Signal calls attention, or may determine that the danger is overestimated or past.*

This red flag or red light (the Cautionary Signal) is only to be displayed when the information in the possession of the office leads to the belief that dangerous winds are approaching.

The term dangerous winds has ordinarily a somewhat different meaning according to the location of the station. Thus the severe gales of the Atlantic (where the hourly velocity of the wind ranges from 40 to 70 miles) are comparatively very rare on the lakes, where the limited sea-room causes winds that on the neighboring shores are registered only as brisk (*i. e.*, 20 to 25 miles) to become dangerous. Again, the direction in which the wind is blowing is a most important consideration, and as general experience shows that most danger is apprehended from wind blowing on to a lee shore, the Cautionary Signal may very properly be expected to be hoisted only in case such winds are apprehended for the port in question.

For inland and well-sheltered points, however, as Baltimore

and Philadelphia, this distinction cannot be easily made, and in order to avoid the confusion that might very possibly arise from the display of different signals at adjacents ports, such as Milwaukee and Grand Haven, Detroit and Toledo, etc., etc., it has been decided for the present, at least, not to put into practice the above suggestion. The Cautionary Signal will therefore be hoisted whenever the winds are expected to be as strong as twenty-five miles an hour, and to continue so for several hours within a radius of one hundred miles of the station. It will thus be left to the public individually to decide whether that wind will be dangerous to any special occupation. It is hoped that eventually it will be practicable to add a second signal, giving warning of severe gales. Each signal holds good for the space of about eight hours from the time at which it is hoisted. When no signal is displayed it indicates that the office has no knowledge of any approaching danger, and as this is not only the case when there is really no danger, but also in many cases may be the consequence of the failure of the telegraphic connection of the Central Office at Washington with neighboring stations, it should not lead the mariner to be less watchful of the weather, nor to neglect to obtain such weather intelligence as he can from the telegraphic reports at the observer's office. If the mariner desires more exact information as to the nature of the threatening danger, he should obtain the latest " Weather-Bulletin," or "Weather-Map," published by the office, as well as the general "Synopses and Probabilities," or the so-called "Press Report." These can generally be had at the office of the observer.

If he finds that high winds are prevailing within two hundred miles of the port at which he is, he should consider what the disturbing cause is that produces these winds. This may in general be stated to be an excess of barometric pressure over some district and a deficiency over some neighboring region.

If the region of low barometer be very small, the area of violent winds will be correspondingly contracted, as in tornadoes on land. Even if no fall in the barometer be noticed, brisk winds may be experienced, owing to the fact that the air in rapid motion overhead may drag along with it that on the surface of the ground, but in general it may be stated that ninety per cent. of the winds that are dangerous to navigators are accompanied by areas of notably high and low barometer. Now, when the barometer falls over any region, the inertia of the surrounding air causes some time to elapse before it is set in motion, and similarly a large mass of air moving with rapidity preserves its motion after the exciting cause is removed. Thus it may happen that strong winds exist in regions at which no barometric disturbance exists at the moment, but has existed a short time previously.

Again, the space inclosing the partial vacuum, into which the wind tends to rush, itself moves slowly over the earth, and thus the wind at any point appears still longer to delay to follow the barometric disturbance. This delay will, of course, vary with the motion of the central area of low pressure, or that of the neighboring high pressure.

The general consequence of the preceding considerations is that the area covered by the weather-chart presents to our view one, two, or three regions of low pressure, and one or two of high, and that between these, but much nearer to the low than the high barometer, we find the strongest winds. As regards the direction of the winds they may be described as not tending directly to the center of the area of low pressure, but as circulating around and in upon it in a sinuous spiral in a direction contrary to the movements of the hands of a watch, (when it is laid down with its face upwards;) thus there are found northerly winds on the west side of the region, westerly ones on the south side, and so around. Of these winds, those from the north-west and south-west are more violent, on the average, than those from the south-east and north-east, but the latter may be more dangerous, and when they pass over smoother ground may be even stronger at the immediate surface of the earth.

The general tri-daily press report (to be had at observers' offices) contains always a statement of the positions and movements of the larger areas of high and low barometer, of cold and warm weather, or of stormy, cloudy, and clear weather. In the absence of a weather map, therefore, one can determine in a general way whether these are approaching to or departing from his neighborhood, and this knowledge leads to the following conclusions:

1. If the barometer is very low at the center, very severe gales may be expected over a large area, say within a circle of two hundred miles radius, from October to April; but within a smaller circle, less than one hundred miles from May to September.

2. Areas of low barometer when first perceived in Minnesota may be expected to move eastward in the summer months with westerly winds on Lake Superior, and to move to the south-east and east-south-east in the fall, with east and north-east winds on Superior and Michigan.

3. When perceived in Nebraska or Indian Territory they may be expected to move northeast to Lake Ontario, with northeast winds on Superior, Michigan, and Huron.

4. When perceived in Texas or anywhere on the Gulf coast they may be expected to move northward to the latitude of 35° or 40°, and there begin to move northeastward, with northeast, north, and northwest winds on the lakes, and, subsequently, southerly winds on the Middle and East Atlantic coasts.

5. When perceived on the coast of Florida, or off the South Atlantic, they, in the fall, winter, and spring, may be expected to move slowly up the coast, preceded by northeasterly winds and rain.

6. Whilst the preceding sentences mark out the most general average phenomena attending the movement of the disturbing areas of low pressure, it must be borne in mind that but very rarely will the ever-varying atmospherical condition allow any storm to pursue a uniform average course over the earth.

7. The most important condition disturbing the regular move-ment of a storm is the presence of moist air; that is to say, air nearly saturated with aqueous vapor. Such air is found over extensive forests and marshes, over bodies of warm water, and especially over a field of snow or ice, which is being melted by the sun. Toward these regions the areas of low barometer may be expected to be drawn more or less strongly, or at least to spread out over them.

8. When the low barometer on the west of the Appalachian chain of mountains causes easterly winds on the Atlantic coast, these force a mass of air up the eastern slopes of these hills, and this, as it rises, becomes more and more nearly saturated, so that the low pressure west of the mountains may be expected to be rapidly spread over the range, and even transferred, as it were, to the Atlantic coast.

9. Waves of low pressure on the Pacific coast are also trans-ferred across the Rocky Mountains and the elevated plateaus, and may be expected, when they arrive at the Mississippi Valley, to produce cloud and, probably, storm centers.

10. Besides the disturbing action of moist areas, we

have to consider the contrary influences exerted by areas of dry air, such as are found on the eastern slope of the Rocky Mountains. When a storm-center is driven into such a region, clouds and weaker winds may be expected, and the storm will gradually die out, unless moist air lies beyond.

11. Besides the forces attracting the areas of low pressure, we have also to consider the presence of any area of high barometer; this, by causing a system of decidedly strong winds to circulate around it in the anti-cyclonic direction, may drive the storm-center (whether it be a small tornado, or an extensive cyclone) before it, so as to undergo quite a change in its path.

12. The general path pursued, day after day, by a storm-center, is, indeed, the resultant of a system of pressures, some of which arise from the pressures of areas of high and low barometer, some from areas of moist or heated air, some from winds, and others from the rotation of the earth on its axis. These are the important controlling influences; such others as may arise from lunar tides in the air, &c., are considered at present as of inferior prominence.

13. Whether the observer determines the probable movement of a given storm-center by means of these general considerations or not, he cannot safely neglect the indications of his own barometer. The experience of the past fifty years has borne uniform, testimony to the pre-eminent value of the indications of this instrument, taken in conjunction, of course, with the wind and weather.

14. If it be known that a center of low pressure is in the neighborhood of the observer, and he stand facing it, he will find the wind blowing from some point on his left toward some point on his right: and conversely if he stand with his left hand toward the direction from which the wind comes, he will face the region of lowest pressure. It is better to use, not the direction of the local wind, but that from which the *low* clouds are moving; the *very high* clouds should not be used.

15. If, while facing the low pressure, he finds the wind steady, and his barometer falling, then the central area is advancing directly toward him, and, so long as this continues, he may expect the wind to increase until the barometer reaches its lowest; then a lull will take place, followed by strong winds from the opposite quarter, which will continue while the barometer is rapidly rising, but subside as it rises more slowly.

The strength of the wind may be expected to be in general proportioned to the rapidity of the fall or rise of the barometer.

16. The storm-center has been spoken of as being in front of the observer; it will, however, in general be somewhat more to his right, as he stands with his left to the wind.

17. If the wind *veer*, (that is to say, gradually change its direction, for instance, from southeast to southwest, or, in that order, around the compass,) the storm-center is passing by on the north side of the observer; if it *backs*, (changing, for instance, from southwest to southeast,) the center of lowest pressure is passing by on the south side, and the distance at which it is may be roughly estimated by the rapidity of the fall in the barometer.

18. The severest winds are those, as before stated, on the south and west sides of the center; those on the east side are more frequently squally.

19. In the foregoing the observer has been supposed to be stationary; should he be in motion, however, this will affect especially his observation of the barometer, since he may easily run into or away from an area of low pressure, and he himself must mentally allow for the influence of his motion.

20. In comparing the readings of his own barometer with those published on the Bulletin and Map of this Office, the mariner will find that the latter have all been reduced to a uniform sea-level, and to a uniform temperature. He, therefore, will find it most convenient to bear in mind the amount of these corrections, and to mentally apply them to his own barometer; should he preserve his records, and transmit them to this Office, however, the original uncorrected observations only are desired. The following tables show the general amount of these corrections.

21. An allowance for the temperature of the barometer is made by noting the thermometer attached to that instrument.

Inch.
For a temperature of 20° Fahrenheit add 0.02
For a temperature of 30° Fahrenheit add 0.00
For a temperature of 40° Fahrenheit subtract 0.03
For a temperature of 50° Fahrenheit subtract 0.06
For a temperature of 60° Fahrenheit subtract 0.08
For a temperature of 70° Fahrenheit subtract 0.11
For a temperature of 80° Fahrenheit subtract 0.13
For a temperature of 90° Fahrenheit subtract 0.16
For a temperature of 100° Fahrenheit subtract 0.18

These numbers are only applicable to barometers having brass scales, which are the most reliable ones; aneroid barometers ought not to require any temperature correction.

The correction for altitude above the sea will vary with the annual temperature, but may be approximately made as follows:

	Altitude of water. Feet.	Altitude of bar. Feet.	Add to bar. Inch.
For the ocean	0	25	0.02
For Lake Champlain	95	120	0.14
For Lake Ontario	235	260	0.29
For Lake Erie	564	590	0.66
For Lake St. Clair	570	595	0.66
For Lake Huron	574	600	0.68
For Lake Michigan	574	600	0.66
For Lake Superior	600	625	0.69
For the Ohio at Pittsburgh	725	750	0.80
For the Ohio at Cincinnati	450	475	0.50
For the Ohio at Louisville	400	425	0.47
For the Ohio at Cairo	300	325	0.34
For the Missouri at Omaha	950	975	1.05
For the Mississippi at St. Paul	650	675	0.74
For the Mississippi at St. Louis	425	450	0.48
For the Mississippi at Vicksburg	130	155	0.17
For the Mississippi at New Orleans	20	50	0.05

PLACING AND READING OF THE INSTRUMENTS.

NOTE.—The following instructions apply to Green's, Fortin's, and other barometers constructed on the Fortin principle, and Robinson's anemometer as constructed by Green, of New York.

BAROMETER.

The barometer must be kept in a room of as uniform temperature as practicable; and to protect the instrument from such external influences as would produce irregularities it should be kept in a box. The box should be firmly attached against the wall in a vertical position, in such a way that when open the barometer may hang in front of a window.

An opening, large enough to admit the tube of the instrument, should be cut in the upper end of the box, and directly above this a strong hook of such length as to extend two or three inches beyond the box, be driven into the wall.

The instrument is to be suspended on the hook, and when not in use to be kept in the closed box.

When an observation is to be made, the barometer must be slipped out on the hook into the full light of the window.

It is always well to follow a system in every mechanical operation, and particularly in taking observations, as it insures an accuracy that cannot otherwise be obtained. The following rules are therefore presented:

1st. Tap the instrument a little above the cistern, to destroy the adhesion of the metal to the glass.

2d. Read the attached thermometer, which is very sensitive.

3d. By means of the adjusting screw bring the surface of the mercury in the cistern in contact with the ivory point which denotes its constant level. If correctly done neither a line of light can be seen between the point and the surface of the mercury, nor will there appear on the surface of the mercury a dimple caused by capillary action.

4th. Again tap the instrument just above the cistern.

5th. Take hold of the instrument above the thermometer, with the left hand, and by means of the vernier screw bring the back and front lines of the vernier into the same horizontal plane with the top of the mercury in the tube, just touching it and no more. Remove the hand; and as soon as the barometer is vertical note whether any line of light appears between the summit and the edge of the ring. When correctly adjusted a small portion is obscured, while the light is seen on both sides.

6th. Read the barometer at leisure, in the following manner:

On the barometer tube is a fixed scale, divided into inches and tenths of inches. There is also a vernier, or sliding scale, which reads to hundredths of an inch.

First read the point marked on the fixed scale, by the bottom of the vernier, which will give the inches and tenths of inches; set this down and then refer to the vernier for the hundredths.

The vernier is divided into ten equal parts, numbered upward from 1 to 10. Commencing at the bottom, examine the lines until one is found exactly coinciding with any line on the fixed scale, the number of such line on the vernier gives you the hundredths, i. e., if the eighth line on the vernier coincides exactly with any line of the fixed scale the reading is .08 inches. In case no line of the vernier exactly coincides with a line on the fixed scale, two lines of the vernier must somewhere be embraced in the space indicated by two successive lines on the fixed scale, and observing where this occurs, read for hundredths the vernier line which most nearly coincides with one of them. In case the coinciding line is 10, which only happens when the zero also coincides, there are no hundredths, and zero must be placed for the hundredths.

The barometrical readings must be reduced for temperature and elevation.

Tables for such corrections can be found in the "Physical and Meteorological Tables," published by the Smithsonian Institution.

This information may also be obtained by application to the observer sergeant in charge of the station nearest to the applicant.

Whenever practicable compare the barometer with any other good one that may be accessible, by making simultaneous readings of both, and preserve the record of the comparison.

THE THERMOMETER.

Place the thermometer in the open air, so situated that it will be always in the shade, and yet have a free circulation of air around it.

The thermometer should be at least from nine to twelve inches from any neighboring object, and should be protected against its own radiation to the sky and earth, and from the heat reflected by neighboring objects.

These conditions can be fulfilled by the construction of an instrument shelter, which may be constructed outside of a window of a room not heated, and which, corresponding in size to the window, should project about two feet from the panes. Lattice blinds should form the exterior of the shelter, these should always be closed as a shelter to the instruments against all radiation, and should be opened only a little in order to admit light when reading the thermometer.

A foot from the panes, and at the height of the observer's eye, two parallel transverse wooden bars about an inch wide should be fastened. The thermometer should be fastened exactly perpendicularly to the bars, so that its top is secured by a screw to the upper bar, while its bulb projects a few inches below the lower bar, to which the instrument is secured by a clasp or screw.

The bulb should be so placed that it will not rest against a wooden or metal back, but be free from both scale and back.

READING.

In reading it is very important that the observer's eye should be exactly at the same height as the top of the column of mercury, otherwise an erroneous reading will be made.

The reading may be best made through the panes, to avoid the influence of the temperature of the chamber on the thermometer, and a second one should be made shortly after to verify the first. When the bulb becomes moistened by rain or fog, or is covered by ice or snow, it should be carefully wiped, and the reading should not be made until the instrument has acquired the temperature of air.

VERIFICATION.

The zero point should be verified unless the thermometer is known to be correct. To do this, immerse the bulb in a vessel filled with snow or pounded ice, and press slightly a layer of several inches around it, so that the stem, which should be exactly perpendicular, is covered with snow as high as the freezing-point on the scale. Do this in a room the temperature of which is above the freezing point, as that point indicates the temperature of *melting* snow.

After about half an hour read it, taking care to have the eye exactly perpendicular to the column of the mercury, and stirring the thermometer about freely in the mixture.

In case the summit of the mercury and the freezing point of the scale do not agree, note the difference. Some instruments are so constructed as to admit of loosening the screws and sliding the glass tube containing the mercury up or down a distance equivalent to the error, but it is not advisable to make frequent mechanical changes of this kind. The correction should be applied to each reading.

SELF-REGISTERING THERMOMETERS.

The two thermometers—maximum and minimum—are to be placed beside the common thermometer, with their bulbs opposite and free, attached horizontally to two perpendicular wooden bars, uniting the parallel bars running across the shelter.

In reading them the same care must be used as with the common thermometer, the eye being in a perpendicular line with the extremity of the index. After verifying the first reading by a second, bring the index of each to the summit of its column by the use of a magnet, in order to set them for the next day's record.

VERIFICATION.

Compare the two thermometers frequently with the common thermometer, and verify the zero several times each year in the same manner as stated for the common thermometer, and enter the error in the register to be applied at each reading.

HYGROMETER.

These thermometers—one with a dry and one with a wet bulb—must be placed on the same parallel bars as the common thermometer and several inches apart. The bulbs should be free and at a distance from the bars.

The cloth covering the bulb should be muslin and of fine texture, and must be changed every month, and the bulb cleaned. It can be washed without removing, by means of a syringe. It may be kept continually wet, or be moistened a short time before taking the observation, and experience has shown that the average result is the same in both cases. Filtered rain-water must be used.

VERIFICATION.

The two thermometers must be frequently compared, and if they are not adjusted so as to correct any difference which may exist, the error must be registered and taken into account after making an observation.

THE ANEMOMETER.

The anemometer should be carefully fixed in a vertical position, upon a post of sufficient height to bring the dial on a level with the eye of the observer, and in an exposed condition so as to receive the full force of the wind. The post should be firmly enough planted to prevent the instrument from vibrating.

To obtain the velocity of the wind at any time, two observations, at an interval of exactly five minutes, should be made, and the difference between the readings, which will be obtained in miles and tenths of miles, multiplied by 12 gives the velocity per hour. Example: Suppose the outer index to be at 3 the first reading, and at 3.6 the second, the difference is 0.6, which multiplied by 12 gives 7.2 miles as the velocity per hour. Great care should be exercised to make these observations exactly five minutes apart.

Reading: Each line on the inner dial indicates 10 miles, and the dial reads by tens from ten to one thousand. Each line on the outer dial indicates a tenth of a mile, and the dial reads, by tenths and by miles, from one-tenth of a mile to ten miles. The zero line of the outer dial is the point at which the inner dial must be read. Read on the inner dial the line exactly coinciding with the *zero line of the outer dial*, or if no line exactly coincides, then read the line next less than it.

No line of the inner dial can exactly coincide with the zero of the outer dial, unless that zero exactly coincides with the steel index at the top of the dials, except when the instrument is improperly adjusted.

When such coincidence does not take place, the outer dial must be read at the point exactly coinciding with the steel index, and the distance there indicated, which is noted on the outer dial in miles and tenths of miles, must be added to the result obtained from the inner dial.

RAIN-GAUGE.

The rain-guage should be placed with the top of the collector twelve inches above the surface of the ground, and be firmly fixed in a vertical position. It should be examined each morning at the usual time of observation, and its contents carefully measured by a graduated rod, which is furnished with the gauge. Snow should be melted, and measured as rain. The gauge should be emptied for each observation. When possible, it is important to keep several rain-gauges in different but adjacent localities, as the results are liable to be much affected by local peculiarities.

www.ingramcontent.com/pod-product-compliance
Lightning Source LLC
Chambersburg PA
CBHW021957190326
41519CB00009B/1300